PQR Theory:
How One Can
Create a Universe

N. H. G. MITCHELL

To John Horton Conway

who taught me about Ford circles

and for whom all numbers are games.

L'universo ... è scritto in lingua matematica, e i carat-
teri son triangoli, cerchi e altre figure geometriche.
(The universe is written in the language of mathematics, and its characters are triangles, circles and other
geometrical figures.)
—Galileo Galilei (1623)

TABLE OF CONTENTS

ACKNOWLEDGMENTS

I extend my thanks to the following individuals who helped me develop this book from a mere idea into a physical reality:

Michael Gotlieb and Elliot Dorff, who provided valuable validation in the earliest stages of my work; Michael and Kayla Stevenson of TransformDestiny.com, whose workshop *The Power to Create Your Book Now!* got me started on writing; my wife Dorine, for her forbearance, support and encouragement throughout the creative process; and my daughter Stephanie, of Stephanie Mitchell's Wordwork (smwordwork.com), for her professional services in proof-reading and editing the first draft.

In the course of my research and writing I made extensive use of several open-source projects, including Wikipedia, FreeBASIC and OpenOffice. Special thanks are due to the volunteers who collaborated to create and maintain these invaluable free resources.

A number of people read the advance draft and made helpful comments and suggestions which I gratefully acknowledge, including Wayne Mularz, Mark Conkle, Sylvie Kern, Don Cohen, Petra Wingert, Felix Barber, Bernard Bronstein, Mary Serdula, Jack De Merit, John Rallison, Ben Mitchell, Andrea Jussim and David Kelley. Any errors and omissions, however, are my sole responsibility.

AUTHOR'S NOTE: A WARNING TO HOBBITS

They like to have books filled with things that they already know, set out fair and square with no contradictions.
—J. R. R. Tolkien (1954)

This book is not filled with facts you already know. On the contrary, it presents a new theory that will change the way you think about our Universe and your place in it. And because the theory is new, I should warn you that not all the details are pinned down yet, and some of them may be contradictory.

That's partly because I decided to write this book before completing my research into PQR Theory—otherwise I would never have written it. And it's partly because *existence itself* seems to be full of contradictions, as you will understand after reading this book.

Sorry, hobbits.

INTRODUCTION
(WHICH YOU SHOULD ACTUALLY READ)

*There is a theory which states that if anyone discovers exactly what the Universe is for and why it is here, it will instantly dis-appear and be replaced by something even more bizarrely inexplicable. There is another theory which states that **this has already happened**.*
—Douglas Adams (1978)

About this book

This book is in two parts. The first part describes the basic concept behind PQR Theory: the idea that our Universe may be built out of num-bers. The second part describes how a particular series of related mathematical models can be used to build structures which appear to resemble our Universe in many ways. There is a glossary after Part II.

This second part is still very much a work in progress. I have not yet shown the resemblance to our Universe to be complete, and I do not yet claim that any of the structures I have created are an accurate model of our Universe. Some of the details may well need to be changed, and indeed the models themselves may need to be replaced by different ones. But I believe that I have now made enough progress to show that many of the principles underlying PQR Theory also underlie the workings of our Universe, and also to illustrate how you can build a whole universe from numbers, starting with just the number 1. Hence the title of this book.

In an attempt to keep this slightly more readable than it might other-wise have been, I have tried as far as possible to keep the mathematics simple and to confine it to the later chapters. For similar reasons, I have tended to be informative rather than rigorous and to provide explanations rather than proofs. This approach may well be unsatisfactory to mathe-matical purists, to whom I apologize, hoping that I may be permitted to remedy this deficiency someday. And the principle of conservation of energy has led me to refrain from explaining some tangential topics that I merely mention in passing. (If you're curious to learn more, you can treat them as references for further reading; Wikipedia is generally a good place to start.)

How this book originated

I hate to disappoint the reader, but there was no single "eureka" moment in the development of PQR Theory. It was more like solving a jigsaw puzzle, with hundreds of little pieces fitting together to gradually reveal the big picture. Sometimes the pieces that seemed to fit turned out to be wrong, and more than once I had to break them apart and start again. But the original seed for this book was planted back in the year 2000, when I read Brian Greene's *The Elegant Universe* with its highly interesting discussion of superstring theory. He explained how the theory had been developed by physicists who had created extra dimensions of space until they had enough of them to build a coherent model that did not conflict with reality. This rang alarm bells in my head for a couple of reasons. First, these extra dimensions of space were so small as to be undetectable and (as I discuss in Chapter 3) I have a deep distrust of physical theories that depend on the undetectable. And second, the development process was like fitting a mathematical curve to a finite set of known data points—the more intricate you make your formula, the closer you can fit your data. By adding enough terms to your formula, you will always be able to match your data exactly, even if it was really generated by a different type of formula—or even if it was totally random. Likewise, by adding enough dimensions to a model of space-time, you can incorporate all known physical processes, even if they might really have much simpler explanations.

On finishing Greene's book, I was left with the feeling that physicists had fallen into this trap of "overfitting the data" and produced a model which is extremely powerful but unnecessarily complex. In other words, superstring theory is too general to be useful—it explains everything but predicts nothing.

But I did glean some useful ideas from Brian Greene. In particular, the extra dimensions involved in superstring theory are so small that they may be limited to just a few bits of information, and some quantum properties seem to consist of pure information. Indeed, I soon realized that the "quantum uncertainty" that physicists observe at the subatomic level could well be the result of trying to extract more information from a system than it contains. For example, measuring the direction of a single electron's spin or a single photon's polarization can produce only a single bit of information; when physicists try to infer any more, they just become confused.

These thoughts led me to reflect on the nature of information, and how it might relate to matter and energy. I started reading more about physics and learned about the Pauli exclusion principle, which says that no two electrons in an atom can share the same quantum state. I also learned how each quantum state can be represented by a set of numbers. Thus, there are a limited number of possible quantum states, each of which would either be "on" (occupied by an electron) or "off" (unoccupied), very much like the binary digits ("bits") of a computer's memory. This rang a different sort of bell in my head: I realized that an atom could be considered as a repository for information. Furthermore, a finite quantity of information seemed to specify the atom's state completely.

This led me to wonder about the following question:

Could matter be comprised of pure information?

If so, that led me down the following chain of reasoning:
1. Cosmologists tell us that the Universe contains only a finite amount of matter;
2. Physicists tell us that all matter consists of finite particles; and
3. Quantum mechanics tells us that each particle only contains a finite amount of information.

Conclusion:
The Universe only contains a finite amount of information!

Since (as I discuss in Chapter 1) any finite amount of information can in principle be represented as a whole number, I wondered whether the Universe itself could be defined by such a number. In other words, could the precise state of the Universe at a given instant be completely specified by a single whole number? Obviously the number would have to be ENORMOUS to hold so much information, but it would still be finite.

Now the space-time in such a universe would not be continuous. (As I discuss in Chapter 2, you need infinite information in order to specify a point in a continuous space.) I therefore began to wonder if the *entire history* of the Universe could be specified as a sequence of these huge numbers: each number in the sequence would define the state of the Universe at a given instant of time, and their progression would define the evolution of the Universe from one moment to the next. This progression could in turn be defined by a formula that encoded the laws of physics. Knowing the formula, one could (in principle) program a computer to calculate the entire history of the Universe—past, present and future!

In the end I realized this was not practicable, as it would require a computer larger than the Universe to calculate this progression of numbers from one moment to the next. So, if the Universe was in fact defined by a progression of numbers, it would have to be a progression that existed naturally, without needing a computer to calculate it. And such a progression does exist: the Natural numbers 1, 2, 3 and so on. I had previously considered these numbers to be too simple and orderly to define anything as complex and chaotic as our Universe, but then I came to realize that they did contain some quite intricate structures. One such structure is the primes—those numbers which can't be formed by multiplying two others together: 2, 3, 5, 7, 11 and so on.

I also realized that my previous idea had been too complicated. I didn't have to describe the entire Universe using just one number; that was asking the number to do too much work. All I had to do was find a formulation that would allow each Natural number to help describe the operation of a tiny little bit of the Universe for a tiny instant of time. Somehow, that task seemed much more manageable for an individual number. And yet it would still allow the series of Natural numbers (1, 2, 3, ...) to define the entire history of the Universe. The smallest numbers would help describe the earliest moments of time, and as the numbers got larger, the universe they described would get bigger too. Of course, our Universe is now billions of years old and has grown to be huge, and the numbers that describe it have also grown to be absolutely ENORMOUS —in fact, so large that to call them astronomical would be an insult. (You may think a googol (10^{100}) is big, but by now the Universe is using numbers *much* bigger than that.) Fortunately the Natural numbers go on forever, so we're in no danger of running out of them any time soon. (By the way, in this book I have capitalized certain adjectives like "Natural," "Rational," and "Real," in order to emphasize that these are *names* for various types of numbers but not necessarily *descriptions* of them.)

By 2008, although I had not yet been able to find a formulation, my ideas had crystallized to the point where I was able to discuss them in a more-or-less coherent manner. My website *pqrtheory.com* went live in

June of that year, and by July of 2010 I had come up with the ideas for formulating the models discussed in the later chapters of this book.

Meanwhile I had started my blog (the link can be found on *pqrtheory.com*), in which I described some of the interesting things to come out of my researches. Some of these ideas I later abandoned, but I have left them up on the blog in case anyone cares to read them.

Much of the material in this book appeared in an earlier form on my website or my blog, but this is the first time I have attempted to collect it together in one place and present it in what I hope will form a cohesive story.

And it's a story that's not finished. My researches continue, and I hope to present more of this story in future editions of this book.

Finally, please bear in mind that not all my parenthetical remarks are intended seriously. As one nineteenth-century mathematician prefaced his treatise on geometry:

I have not thought it necessary to maintain throughout the gravity of style which scientific writers usually affect.

Elsewhere, that writer questioned the use of a book without pictures or conversations. This book (being pre-Socratic in its philosophy) may lack conversations—but hopefully it has enough pictures to make it of some use.

N. H. G. Mitchell, November 2017.

PART I: PQR THEORY

1. What is PQR Theory?

Nil igitur fieri de nilo posse fatendumst, semine quando opus est rebus. (We cannot conceive of matter being formed of nothing, since things require a seed to start from.)
—Lucretius (first century BC)

PQR Theory is a new theory about the nature of our Universe. It seeks to explain how the physical phenomena of matter, energy, space and time can all be formed from numbers.

It is so named because it combines three different lines of thinking: first, the teachings of Pythagoras that "All is number"; second, quantum physics, which states that energy always exists in little packets; and third, Einstein's relativity theory, holding that matter and energy are equivalent and that space and time are interconnected. Hence the name Pythagorean Quantum Relativity, or PQR for short.

PQR Theory takes these principles a step further by holding that matter, energy, space and time are all different aspects of one single structure, built from the numbers 1, 2, 3 and so on. In other words, *our Universe is made from numbers*.

Our Universe: Matter, energy, space and time

At the end of the nineteenth century, physicists thought they had a definitive model of the processes which operate our Universe. There was matter, the stuff of which everything was made; energy, which gave matter heat, or height, or motion; space, the unchanging three-dimensional background against which everything happened and against which everything could be measured; and time, the universal clock which told you the order in which events occurred. Matter was understood to consist of indivisible particles called atoms (meaning "uncuttable"), but space, time and energy were believed to be continuous—they could be divided and subdivided indefinitely, into parts as small as one wished. In other words, you could never have less than one atom of matter, but for space, time or energy there were no minimum amounts: however tiny a quantity you took, you could always divide it into two halves.

These four fundamental components of the Universe were thought to be independent and immutable: matter and energy could not be created or destroyed, only transformed; space and time were not influenced by each other or by anything else. Energy was something that inhabited matter, and at any given point in time, each particle of matter occupied a definite position in space. A particle's "motion" reflected how this position changed with the passage of time; a "force" was something that caused it to change its motion; and energy could be converted into, or derived from, a force acting on matter as it moved through a distance. Matter and energy were said to be *conserved,* meaning that the physical processes occurring within any closed system could never change its total matter or its total energy. This was the generally accepted classical or Newtonian model of the Universe prior to 1900.

Of course physicists knew about phenomena like gravity, light, electricity and magnetism, but these were all explained in terms of the four fundamental components. For example, gravity was a force that acted on matter; light consisted of particles, or possibly waves, moving through space; and electricity was a fluid that exerted a force on matter. Physicists thought they had a complete and consistent theory, as illustrated by the following (probably apocryphal) quote attributed to William Thomson, Lord Kelvin in 1900:

There is nothing new to be discovered in physics now. All that remains is more and more precise measurement.

Relativity and the quantum theory

The classical model agreed well with the experimental observations that were possible using Victorian technology: any discrepancies seemed to lie within the limits of experimental accuracy. But with the dawn of the twentieth century, there came a growing realization that this model was inaccurate at very small scales, and also at very large speeds like that of light.

Then Einstein, attempting to understand and explain some of these discrepancies, came up with his theories of relativity which stood the old assumptions on their head. Matter and energy were actually one and the same thing, and space and time did not exist independently but were different facets of a single four-dimensional structure ("space-time") that was influenced by the presence of matter-energy. Relativity has subsequently proved to be very accurate in modeling how movement and gravity distort space and time. For example, the atomic clocks on board GPS satellites are specially adjusted to allow for relativistic effects. Without these adjustments, the navigation system in your car would "drift" about a quarter of a mile an hour, rendering it useless for all but the shortest journeys. Einstein's relativity theories have also explained, with remarkable accuracy, the minute distortions that astronomers have observed in the light coming from distant stars.

Meanwhile, other scientists probing the atom's internal structure showed that (for any given frequency) energy always came in discrete packets, each containing a particular amount of energy, and that each packet or "quantum" held the smallest possible amount of energy that could exist. A quantum could not be subdivided, so you could never have fractions of a quantum—you could only have a whole number. And these quanta behaved unpredictably: you could never tell exactly where one was going to turn up, although if you had a lot of them, you could accurately predict how they would be distributed.

The resulting quantum theory models the behavior of matter and energy at the subatomic level, where these two concepts tend to merge. At these tiny scales, particles of matter and waves of energy become indistinguishable. Sometimes it's more convenient to treat them as particles and sometimes as waves, but we should always bear in mind that these are just different ways of viewing the same underlying quantum of "matter/energy." And these quanta do not have definite positions, but instead have to be considered as "smeared-out" waves of probability in space and time. So you can never know exactly where or when an experiment will detect a particle. And the more accurately you try to measure its position, the less accurately you can measure its wavelength. Likewise, the more accurately you try to measure its energy, the less accurately you can know the time when the measurement is made. These uncertainties don't arise from deficiencies in the measuring equipment, but appear to stem from a fundamental "fuzziness" in the nature of matter and energy. Your measurements will always be at least as fuzzy as the quantity you are trying to measure (and to make matters worse, the act of measurement tends to influence what you're measuring). There seems to be a fundamental limit to how much information you can extract from a particle, no matter what quantity you try to measure. Yet despite this "quantum weirdness" and the philosophical difficulties it poses, the quantum theory has proved to be extremely accurate at modeling the inner space of the atom, just as its sister relativity has proved in outer space.

These new branches of physics were initially hard to grasp, as they seemed to overturn the well-understood classical model. But by the time the energy locked within the atom was dramatically unleashed at the end of World War II, the truth behind these theories had become hard to challenge. Of course, they did not so much overturn classical physics as extend it. Quantum theory dealt with events on the smallest scales, while relativity handled events on the largest scales. And in between, on the scale of everyday events, both theories agreed with classical physics.

But there remained a problem. While quantum theory and relativity have each proven very successful in their own fields, and while both are compatible with classical physics, they are not compatible *with each other.* They can't be combined into a single consistent picture of the Universe—the so-called "Theory of Everything" which physicists, mathematicians and philosophers have long sought. In order to bridge the gap between these two theories and harmonize them into a coherent whole, we will need to bring in a third, much older, theory.

Who was Pythagoras?

Pythagoras of Samos
Was moderately famous
His theorem on the hypotenuse
Gotta lotta news.
—Anon., after Edmund "Clerihew" Bentley (1875–1956)

Much has been written but little is known for sure about this charismatic Greek philosopher who lived in the sixth century BC. He is best known today for the geometrical theorem that bears his name, although he probably was not the first to discover it. But he is said to have been the first man to call himself a philosopher, and he founded a school of philosophy (or religion—the two ideas overlapped) that taught a way of living as well as a theory of nature. And although many of his cult's teachings were kept secret and are lost to the ages, one idea that has come down to us is his saying that "all is number."

This concept is at the heart of PQR Theory, although implemented in a way that almost certainly was not a part of ancient Pythagoreanism.

As we have seen, modern physics holds that matter and energy are the same thing, and that they are "quantized," coming in individual, indivisible packets. It also holds that space and time are different aspects of the same four-dimensional structure. PQR Theory combines these lines of thought and states that *all four* of the Universe's fundamental components, matter, energy, space and time, are quantized and that they are all different aspects of a single structure that is formed from the Natural numbers.

What are the Natural numbers?

Die ganzen Zahlen hat der liebe Gott gemacht, alles andere ist Menschenwerk. (God created the whole numbers, everything else is Man's work.)
—Leopold Kronecker (1886)

Mathematicians recognize and study many types of numbers.

The simplest, most basic type is known as the Natural numbers. These are the counting numbers with which we are all familiar: 1, 2, 3, 4 and so on. And they all stem from the number 1, one, unity, the monad. Armed only with the idea of unity and the concept of adding, you can create all the Natural numbers by starting with 1 and adding 1 over and over again: $1 + 1 = 2$, $2 + 1 = 3$, $3 + 1 = 4$ and so on.

The Natural numbers go on forever or, as mathematicians say, up to infinity. A word of caution here: "infinity" is not really a number, but merely a mathematician's way of saying that numbers continue indefinitely without ever stopping. In other words, if I think of any Natural number, you can always add 1 to it to get a bigger Natural number. There is no "largest" Natural number.

You can add or multiply any two Natural numbers together, and your result will always be a Natural number. But this is not always true when you subtract: sometimes your result is zero or a negative number. Results like these are not included in the Natural numbers.

So, in order to be able to solve their problems, mathematicians need to use other kinds of numbers aside from the Natural ones; but keep in mind that these are artificial creations and do not represent anything real. You can't have minus seven apples.

The first extension to the Natural numbers is known as the integers, or the whole numbers. These consist of the Natural numbers together with their negatives (−1, −2, −3, −4 and so on) and zero. They can be written as the set {..., −4, −3, −2, −1, 0, 1, 2, 3, 4, ...}, or equivalently you could imagine them as points marked out at equal intervals along a number line extending to infinity in both directions.

··· **-4 -3 -2 -1 0 1 2 3 4** ···

Now that we have the integers, we can add, multiply or subtract any two of them and our result will always be another integer. But when we try to *divide* one by another, the answer often turns out to be a fraction rather than a whole number. So for division, we need to supplement the integers by including all possible fractions that can result from dividing one integer by another. (Note that division by zero is forbidden.) This gives us the Rational numbers, so called because each one is the *ratio* of two integers.

The Rational numbers include all possible fractions like ½, −2¾, or 355/113. Note that the integers are themselves Rational numbers (e.g., 6/3 = 2), and that there are many ways of obtaining each Rational number. For example, 1½ can be written as 3/2 or 6/4 or −300/−200, and these three division sums all result in the same Rational number.

We can regard the Rational numbers as filling in the gaps between the integers on the number line drawn above, just like the markings on a ruler would if they could show *every* fraction of an inch. Now, at last, we can add, subtract, divide or multiply any two Rational numbers (apart from dividing by zero) and the answer will still always be a Rational number.

So far, so good. But there are some problems that mathematicians can't solve using Rational numbers alone, and so they have created other kinds of number. As we will see in the next chapter, the trouble is that these new numbers come at a price that we—and indeed the Universe—cannot afford.

What is information?

To understand the difficulty, we will need to consider the nature of information. But what exactly *is* information? The question is simple to state but subtle: the term refers to various overlapping concepts, and there is no consensus on its precise meaning. For example, if I tell you something you don't know, that would be information. But if I tell you the same thing twice, have I given you any more information than by saying it just once? And if I then tell you something to the contrary, what does that do to the information you already have?

So to reduce the confusion, we should first note that information is not the same thing as *meaning*. Meaning relates to how we might use a piece of information. A string of random digits by itself contains information but has no meaning. But if you also know that these digits represent the combination to a safe, that information becomes meaningful.

Fortunately, we are not concerned with meaning here, just with information as an abstract concept. But this raises a deep philosophical question: *can information exist in the abstract?* For example, consider the mathematical constant π (pi, the ratio of a circle's circumference to its diameter). This is *approximately* 3.14159, but it cannot be written down exactly, as its digits go on forever. Now one could argue that the hundredth digit of π has always existed as a theoretical abstraction; but before anyone actually calculated it, was it information? Maybe it existed in some ethereal realm of higher mathematics, but did it exist in our Universe before someone first wrote it down? (It was the aptly named John "calculating" Machin, in 1706—the same year that the symbol π was introduced.) By now of course, some computer in Switzerland with nothing better to do has calculated π to more than *twenty-two trillion digits* (which didn't leave its hard drives with much room for *useful* information). But do the remaining, uncalculated digits yet exist?

I would argue that the answer is no (at least, not on this planet); in order for information to exist in our Universe, it must be encoded into some physical medium, such as sound waves in air (which only exist for a short time) or ink marks on paper (which last a lot longer). This approach implies that information can be copied, created and destroyed. Our memories are one form of information, but they die with us.

The medium is not the message
In considering this topic, we must take care to distinguish between the physical medium on the one hand and the information that it encodes on the other. The information is not the paper or the ink marks or even the pattern that they form, but rather the actual underlying message that is encoded. For example, words are a widely used form of information. A

word can be spoken as a series of sounds or written as a series of symbols, but it is the same word in whatever medium it is encoded. And although many words have a range of meanings, remember we are not concerned with meaning here: it is the *word itself* that is the information. (Be warned, however, that our English-language encoding system is not perfect: different words can be encoded the same way. For example, the four-letter sequence *lead* can represent several different words, including one that means "to go in advance or direct" and another that means "a heavy metallic element," and the spoken sound *led* suffers from a similar ambiguity.)

Apart from words, many other forms of information are in common daily use. These include numbers, symbols and musical notation, as well as the information contained in sounds, pictures and video. So for a working definition of information I will be pragmatic and say that *information is something you can store in a computer*. (Of course so is a microchip, but let's not be pedantic here.) And you can measure any piece of information in terms of the amount of computer memory it occupies (although this is not an absolute measure, since it depends on how the information was encoded when it was put into the computer).

Conceptually, computers store information as a series of zeros and ones known as binary digits or bits. Each piece of information, whether it be a word, a number, a picture or a movie, is stored as a string of these bits. The longer the string, the more information it contains: a 64-bit string contains exactly twice as much information as a 32-bit string. But every type of information is stored in the same form: a string of bits. (These strings often also include extra "control information" to tell the computer how to decode the data into the appropriate form when it is retrieved from memory.)

So now, in the abstract, we can conceive of information as something that can be represented by a string of zeros and ones. The zeros and ones are the "bits" of information, and this is what raw information is, in its purest form.

We can convert any such string into a number simply by putting a 1 in front of it (so it doesn't begin with zero) and treating the result as a binary number (although potentially a *very* large one if the string is long). For example in the coding system ITA2, the word *dog* would be represented by the string 010011100011010. Put a 1 in front of this string and we get the string 1010011100011010, which is the binary form of the decimal number 42778. Likewise, *cat* becomes 47216. We now have a way in which abstract concepts can be represented by Natural numbers (although most numbers in this system won't have any meaning). By using a system of this type, we can represent *any* piece of information *by a single Natural number.* And conversely, any Natural number will represent a unique string of pure information. So if the Universe itself consists of information, this information can be represented by Natural numbers. Pythagoras, I am sure, would have approved.

What is Pythagoras's theorem really telling us?

Many cheerful facts about the square of the hypotenuse.
—M. G. Stanley (1877)

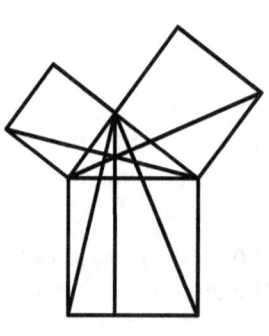

At school, we all learned about the geometrical theorem that says "in any right-angled triangle, the square on the hypotenuse is equal to the sum of the squares on the other two sides." Also known as Euclid's forty-seventh proposition, it is typically illustrated by the diagram at right, accompanying a geometrical proof that the combined area of the two smaller squares is equal to the area of the large one.

From this we infer the formula $a^2 + b^2 = c^2$ relating the lengths of the sides a, b and c of a right-angled triangle, as in the following diagram:

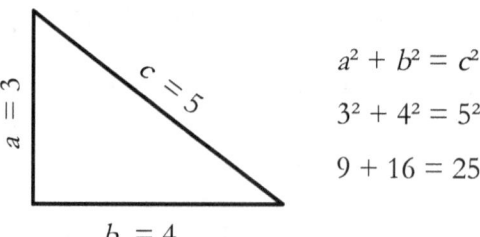

$$a^2 + b^2 = c^2$$
$$3^2 + 4^2 = 5^2$$
$$9 + 16 = 25$$

Wow, we think (if we're still awake), *what a cool fact about right-angled triangles.* But if we stop at that point, we will have missed a deeper truth about distances and space.

For the theorem also works *in reverse*. So we can take any set of three Natural numbers {*a*, *b*, *c*} that satisfy the equation, for example {3, 4, 5} or {5, 12, 13}. (Such a set is known as a "Pythagorean triple.") And now the theorem tells us that any three points in space which are separated by those three distances will define a right-angled triangle. This is a geometrical statement about an angle between the lines joining those points—in other words, Pythagoras is now telling us *how numbers control space itself.* As we will see, this idea is at the very heart of PQR Theory.

2. Physical Existence

The universe may
be as great as they say.
But it wouldn't be missed
if it didn't exist.
—Piet Hein (1968)

What do we mean by the term *existence*? Can we be sure about what exists and what does not?

Well to start with, when something is directly perceptible to our senses, we can be fairly confident that it exists. I know the desk at which I am writing exists because I can see and feel it, I can hear it if I tap on it (and I could probably smell and taste it too). I can also *infer* that the desk exists, because otherwise my computer would be on the floor. This inference is based on my understanding of how the Universe operates: I know from experience that objects like computers do not remain unsupported in midair.

In fact, what we call our sensory perceptions are really also inferences based on our early childhood experiences of how things work. As babies we learn to recognize things like desks and how they affect our senses. This sort of understanding has to be learned: for example the game of peekaboo teaches us that things are still there even when we cannot see them. (Once we grasp this idea, the game loses much of the element of surprise that previously made us laugh.)

Of course the phenomena we observe need to be independently verifiable by others, before we can be sure they exist. I understand that the objects in my dreams *don't* exist, because no one else can see them. And if I were the only person who could see my desk, it would mean there was something extremely fishy going on. (That's why there's no such thing as a headache: it's all in the mind.)

Even if we cannot perceive something directly, we can also be fairly confident it exists if it can be detected by our scientific instruments. But now we're on shakier ground, because our so-called laws of physics are not the absolute laws that control the Universe's operation: they are merely our mathematical descriptions of how we have observed it to work. So our understanding of what exists is based on our models of how the Universe and our instruments operate. For example, we can see Mars with the naked eye or through a telescope, so we infer that the images we see through the telescope reflect reality (albeit upside down). But we must be careful about how we interpret what we see—Mars doesn't have canals, as was once thought (although it may once have had rivers and may still have streams).

Likewise, we say that a compass needle detects something called a magnetic field, but does that field really exist? Or is it just something we have made up, a physical model to explain how compass needles behave? It has proved to be a valuable model, because it has allowed humanity to navigate the Earth and to build useful things like electric motors (as well as less-useful things like Big Mouth Billy Bass). But it is a model nevertheless, and it is not the best model of how things behave at the quantum level. So again we must be careful: instead of treating the field as something real, we should recognize that it's just a formula to predict which way a compass needle will point when we put it some-where. The field is simply our large-scale description of the much smaller underlying causes that combine to make the needle point a cer-tain way.

This is an example of a theme that we will find recurring again and again in PQR Theory: the bits of information which make up our Uni-verse are too small for us to detect directly. We can only observe the larger-scale phenomena that they cause, and then try to infer what is going on at the most fundamental level.

We see this when we try to consider the nature of light. We know that light is made up of things called photons, which are the smallest pos-sible units or "quanta" of light. These photons carry energy and transfer it from one place to another. We have formulas that can model very accu-rately how photons behave, but scientists do not know exactly what they *are*. In some respects they seem to behave like waves; in others, like par-ticles. But neither the wave model nor the particle model provides a complete description of light.

Do photons exist?

Photons presumably do exist, because we can see them—they affect our eyes. But does a photon exist *while it is in transit* from one place to another? If a photon departs from place A at time *a* and arrives at place B at time *b*, can we say anything about what happens in between? From our perspective, it may seem that the photon travels in a straight line from A to B at the speed of light, but careful experiments have shown that we cannot know anything about the path it has taken. This is because we cannot detect the photon at any intermediate place between A and B without stopping it from reaching B.

We can shed some light (pun intended) on this paradox if we consider it from the photon's perspective rather than from our own. A photon travels at the speed of light, and relativity theory tells us that time slows down as you travel faster; if you could travel at the speed of light, time would stop altogether. In other words, if the photon could magically carry a clock along with it, that clock would register *no time at all* passing during the journey. So *from the photon's perspective,* the journey is instantaneous: it leaves A and immediately arrives at B. There is no intermediate "time" when it is between A and B. Thus the concept of its passing through a string of places between A and B is meaningless—it jumps directly from A to B without passing through any intermediate points.

But, you may ask, if this is the case, why can't a photon pass through a solid wall? The answer seems to be that for a photon to pass, there must be some sort of a direct connection from point A at time *a* to point B at time *b*, which a solid wall would not permit. I will discuss this topic in a bit more detail in Part II, but first it is necessary to consider the nature of space and time.

Un-real numbers
Math's just physics unconstrained by precepts of reality.
—Randall Munroe (2012)

The classical view of space and time, and indeed the one that underlies both the quantum and the relativity theories, is that space and time are *continuous*. This means that space and time are treated as being perfectly smooth, with no gaps—not even infinitesimal ones. One can in principle have a distance or a time interval as small as one wants; and one can identify a position in space (or an instant in time) that corresponds to any arbitrary number.

PQR Theory holds this principle to be a fallacy: space and time are not continuous; instead, they consist of a series of separate points. But to understand the reason for this conclusion, we will need to consider what mathematicians very misleadingly call the Real numbers.

We have seen how the set of integers could be extended to include fractional values like 1½ or 22/7, which are called the Rational numbers because each one is the *ratio* of two integers. But mathematicians long ago found that some problems cannot be solved using only the Rational numbers.

For example, consider the equation $x^2 = 2$, which represents the question "What number, multiplied by itself, gives 2?"

Mathematicians use the symbol √2, the square root of two, to denote the answer. It represents the ratio of a square's diagonal to the length of its side, but it's long been known that this is *not* a Rational number. (This fact troubled the ancient Pythagoreans, who tried to cover it up—reputedly by throwing the man who revealed it off a ship.) It's easy to see that √2 lies *somewhere* between 1.41 and 1.42; with more precise calculation you can say it lies somewhere between 1.41421 and 1.41422, but while you can always narrow this gap by calculating more decimal places, you can never pin the answer down precisely, no matter how many digits you calculate.

The reason is that—as was the case with π—the decimal places of √2 go on forever, without repeating or stopping. However, all Rational numbers have decimal expressions that either terminate or eventually go into an endlessly repeating loop. Thus √2, like π, appears to occupy an infinitesimal gap in the sequence of Rational numbers. (Geometrically, if you try to mark all the Rational numbers by placing dots along a line, you will find that however close you place the dots, there will always be tiny gaps between them.) The Rational numbers are thus, in some sense, incomplete.

There are many other problems in mathematics that the Rational numbers can solve only approximately, never exactly. The solutions of these problems all have decimals that go on forever without repeating, which means that they can never be found by taking the ratio of two whole numbers. Numbers like these are known as Irrational numbers and in some sense they represent the infinitesimal gaps between the Rational numbers. I say "in some sense" because there turn out to be *more* Irrational numbers than Rational ones! This is a problem, as we will soon see.

If you are allowed to use Irrational numbers to plug the gaps in the line of Rational numbers, you can create a set called the Real numbers, which consists of the Rational numbers supplemented with the Irrational ones. Geometrically, the Real numbers represent a continuous line with *no* gaps, not even infinitesimal ones. Before I explain why the term "Real" is a misnomer, I will briefly mention yet another class of numbers that mathematicians and physicists often use: the Complex numbers. These are numbers that have been invented in order to solve equations like $x^2 = -1$, which cannot be solved using Real numbers. Mathematicians have invented the "Imaginary" number i to solve this equation; the resulting Complex numbers (which consist of a Real part *plus* an Imaginary part) have proved to be very useful for physicists' calculations. The physicists understand that they are working with imaginary quantities, artificial creations of mathematics, but they are happy to do so because their answers agree well with their experimental results.

But in the case of Real numbers, the problem is that the Irrational numbers (which make up over 99.999999999% of the Real numbers—with as many 9s as you care to write) each *contain infinite information*. A number like π or $\sqrt{2}$, whose decimal digits go on forever, cannot be represented as a finite string of digits, or even as one that repeats. You can never calculate these numbers exactly, because there will always be more digits to calculate. The only way to pin these numbers down would be to use an infinitely long string of digits, which means that a single Irrational number like $\sqrt{2}$ would contain more information than the entire Universe! And while $\sqrt{2}$ is a magnificent shorthand way of writing this number, it refers to a number that *cannot possibly exist* in our finite Universe.

Mathematicians long ago recognized this problem and realized it was insoluble. So they neatly sidestepped it by creating a new axiom that would let them formally prove theorems about the Real numbers. In simple terms, this *axiom of choice* presupposes that you are allowed to make an infinite number of choices in a finite time. Now you can pick infinite sequences of digits in order to create Irrational numbers.

While this axiom is convenient mathematically because it allows answers to problems (like $x^2 = 2$) that would otherwise have no solution, it has no relevance to the physical Universe in which we live. Mathematicians and physicists are apt to overlook this point and have tended to assume that things (like the circumference of a circle) could be measured as accurately as desired, provided one had good enough instruments to measure them with. This assumption is now known to be wrong. The uncertainty principle of quantum mechanics (introduced by Werner Heisenberg in 1927) provides a limit on how accurately things can be measured at any given time. And even ignoring this principle, in practice *any* measurements we make will always be Rational numbers, simply because these are the only ones we can write down.

So the Irrational numbers have no relevance to anything we can measure. Like the Complex numbers, they are simply creatures of mathematics, useful perhaps for performing theoretical calculations, but never appearing as the results of real-life experiments. And that is why the term *Real numbers* is so misleading: most of them are Irrational, so they can never represent anything in our physical Universe.

Is space continuous?

But what, I hear you say, about the circumference of a circle, or the diagonal of a square? Surely these quantities have precise meanings? Even if we set aside the question of whether one can have a perfectly circular or square physical object with *exactly known* dimensions, there is a deeper, more fundamental question: is s*pace itself* continuous? We've long assumed this to be the case, but this tacit assumption must now be seriously questioned. The problem is that in a Universe with finite information, this information can only specify a finite number of positions in space. And since it requires infinite information in order to specify a "Real" number, it would also require infinite information to specify its precise position on the number line. This means that in our physical Universe you can never draw an absolutely smooth line with no gaps—at best you can have a line which looks like a ruler divided into fractions of an inch, a bit like this:

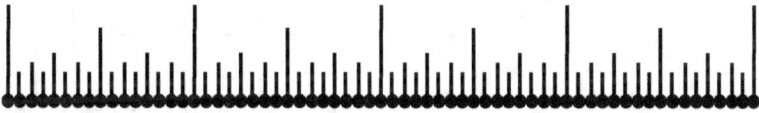

Now, if space itself is considered as a series of dots with nothing between them, then the distance from one point to another can never be measured along a straight line or the arc of a circle. It can only be measured by jumping from one dot to the next and counting how many jumps are made. In this hypothetical scenario, all distances would be an exact number of jumps, and it would be meaningless to talk about fractions of a jump, because *there is nothing in between the dots.* In such a space it is not possible to draw a perfect circle or a perfect square, because all drawing has to be done by "joining the dots."

Such a space may seem very strange, but we are all familiar with this idea from everyday life. Whenever we look at a photograph in a newspaper or an image on a video screen, that image is made up of tiny pixel dots. If you enlarge such an image, you will see ragged edges that previously looked smooth; this is because the space underlying the image was not smooth, but discontinuous. At first glance the picture may have looked smooth, but that was only because the underlying space was finely enough grained to deceive the naked eye.

The space in our Universe is far more finely grained than the space in a picture: the points are *very* close together, so that for all practical purposes physicists can do their calculations as if space and time *were* continuous. But this is really only an approximation (although a very close one, due to the great age of the Universe), as we will see.

In PQR Theory, space-time is considered to be *granular*. This means that at any given time, the fabric of space can be likened to a jar full of coffee granules, each granule representing a single point of space. There is nothing between the granules, yet they fill the jar—any spot you point to in the jar will belong to a particular granule. But there are only a finite number of these granules, which means that you might point to two different spots that you think are close together in space and find that they are both in the same granule. This also means that a granule does not have a single well-defined position in space: if you try to treat space as being continuous, you'll find that each granule appears to occupy a small region rather than being located at a single point.

We can illustrate this with a simple example. These two diagrams

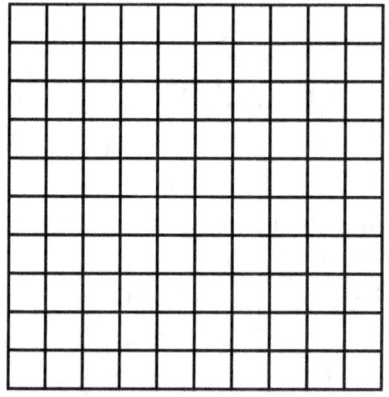

illustrate these two equivalent ways of viewing this scenario. Both diagrams depict the same theoretical two-dimensional granular space, consisting of 100 points arranged in a 10 x 10 grid. The picture on the left shows it as a grid of individual points; the one on the right as a grid of square cells. The first picture shows it as it really is, a bunch of points with nothing between them, while the second shows how it would look if you were to consider it as a continuous space, but they both depict the *same underlying space*.

The death of pi

How might the laws of geometry operate in a world with such a granular space? Imagine a large sheet of paper filled with a grid like either one on the previous page, but with the points or lines spaced one millimeter apart. We are allowed to measure the distance between two points or cells in this space, but *only in whole millimeter units*. We do this by counting the numbers of horizontal and of vertical steps that must be taken to get from one to the other. Calling these two numbers x and y, we can then use Pythagoras's theorem to calculate the "Pythagorean distance" between these two points (or cells). The distance is given by the formula $\sqrt{(x^2 + y^2)}$, but we have to round the result down to the next lower whole number, because fractions of a unit are not allowed. We are also allowed to calculate areas: the area of a region (in what we'll call square millimeters) is found simply by counting the number of points or cells it contains.

Here is how we find the area inside a circle of radius r millimeters. First, we mark one point (or cell) on the paper as the center of the circle. Next, we mark all other points (or cells) whose Pythagorean distance from the center is less than r. Finally, we count the number of marked points (or cells). The diagram on the next page (which has been drawn using cells rather than points, and enlarged for the sake of clarity) illustrates this process by showing two circles with radii $r = 10$ and $r = 20$ millimeters, which turn out to have areas of 305 and 1,245 square millimeters respectively.

Note that we have not used a ruler or compass to mark the regions and find the areas; we did it by counting and arithmetic only. Note also that in this granular space we cannot mark *part* of a cell: it must be marked in its entirety or not at all.

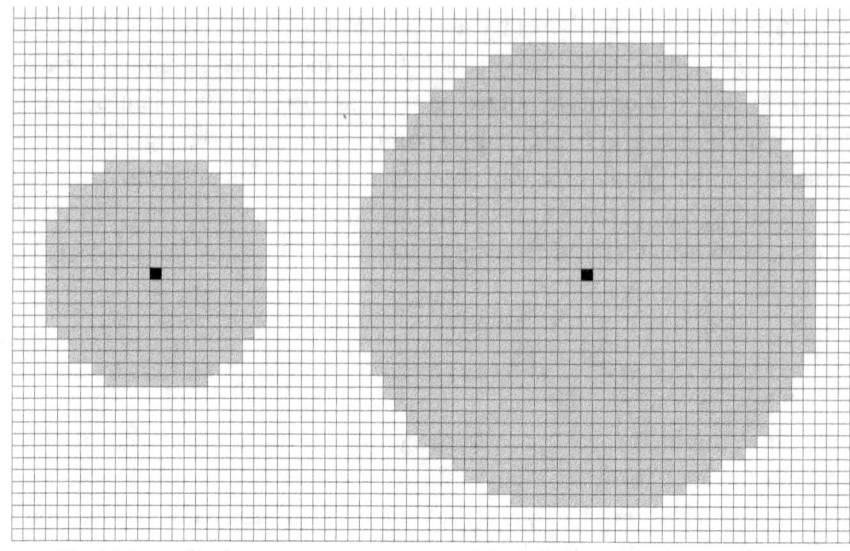

r = 10; 305 marked squares   r = 20; 1,245 marked squares
      305/100 = 3.05   1,245/400 = 3.1125

This process gives us the *exact* area in square millimeters of the (roughly circular) region of the paper that we have marked. This number will be approximately π times r^2, so if we divide the area by r^2 we can obtain an estimate for π. Thus, the two circles in our illustration yield estimates of 3.05 and 3.1125 respectively. (Our estimates will always be Rational numbers, since r always has to be a whole number in this space.) Some estimates for other values of r are shown in this table:

Value of r:	Area of region:	Estimate for π:
5	69	2.76
10	305	3.05
20	1,245	3.1125
50	7,825	3.13
100	31,397	3.1397
200	125,609	3.140225
500	785,321	3.141284
1,000	3,141,521	3.141521
2,000	12,566,317	3.14157925
5,000	78,539,641	3.14158564
10,000	314,159,017	3.14159017

The larger we make our circle (or the finer we make our grid), the closer to a true circle will our marked region tend to be; as the table shows, our estimates for π become more accurate as r increases.

Now imagine that our sheet of paper represents a two-dimensional world in which space is granular, so each dot or cell on the page represents an indivisible unit of space. If there were mathematicians living in this dotty cellular Flatland, they could in principle perform the same calculations as we did and compute π to any degree of accuracy they desired. They would then, like us, be able to say "the area of a circle is the square of its radius times 3.14159265..."

But they would be wrong. In the granular space of this world, *there are no fractional areas—and no perfect circles.* The regions that *look* to be circular in this space are really made up of individual points or cells. And although they could compute the *mathematical* value of π as precisely as they wished, they would realize (if they could see the individual granules of their space) that the area they had marked was only *approximately* equal to πr^2: the *true* area would be as shown in the table.

You may think that this discrepancy arises because the marked region is not perfectly circular, and from our viewpoint this would be true. But to the inhabitants of this world (who cannot see the dots or the grid lines), the region *would* appear to be perfectly circular to the best of their measuring ability. In the same way, our physical laws, which are based on the assumption that space and time are infinitely divisible, can only approximately predict the actual results of any experiment (which of course can only be reported using Rational numbers). Although these approximations may be very close, there will always be experimental errors. Physicists ascribe some of these errors to "quantum effects," which is really just another way of saying that any theory that uses a continuous formula can, at best, only approximate reality. Their formulas assume that we are in an idealized universe whose space is continuous, but according to PQR Theory, their theoretical model only approximates the granular space of our real Universe.

The space-time discontinuum

This granularity of space extends into the time dimension also, although this is harder to visualize. In relativity theory, space-time is considered as a four-dimensional structure, having three dimensions of space and one dimension of time. In order to specify the position of an object in a three-dimensional space, you need three numbers known as *coordinates*. (For example, on the Earth you could use latitude, longitude and elevation as your coordinates—three numbers which specify the position of any point relative to the Earth's surface.) In a four-dimensional space-time, you need an extra coordinate: the time. Taken together, these four numbers will specify the position of an *event* in the space-time. For example, consider the dropping of the Times Square Ball (an oxymoron?) that marked the new millennium in New York. The four coordinates (latitude, longitude, elevation and Julian date/time) for this event could be written approximately as "40.7564° N, 73.9865° W, 480′ elev., JD 2451910.7083." (For greater precision, you could specify the elevation and time separately for the *start* and the *end* of the ball drop, since these are two separate events with different positions in space-time.) The point is that any position in space-time can be uniquely specified by just four numbers, and so we consider time as a fourth dimension in addition to the three dimensions of space.

In normal three-dimensional space, we call a ball's shape a *sphere*, and its surface can be described mathematically by a simple formula involving the three coordinates of space. This formula can be extended to cover the four dimensions of space-time, and now it describes a four-dimensional object known as a *hypersphere*. Now if you have read *Flatland,* Edwin Abbott's book about a two-dimensional world, you may recall how the three-dimensional Sphere appeared to that world's inhabitants: first as a point, then growing into a circle, then shrinking back down to a single point and vanishing. Being creatures of two dimensions, they could only see the circular cross-section of the Sphere as it passed through their plane. Likewise, in our three-dimensional space, an imaginary hypersphere would show the same behavior: we would only see its spherical cross-section. So we can visualize a hypersphere in space-time as a "bubble" that appears as a single point in space, grows rapidly to become a sphere of a certain size, and then equally rapidly shrinks back down to a single point, which then vanishes. The sphere that we see at any particular instant of time is in fact a three-dimensional "slice" of the four-dimensional hypersphere.

In relativity theory, space and time are connected by the speed of light, which is approximately one foot per nanosecond (a billionth of a second). Thus, a hypersphere with a diameter of one foot in the spatial dimensions would have a diameter of one nanosecond in the temporal dimension. In other words, it would last one nanosecond from its appearance to its disappearance, and at its maximum size (halfway through that nanosecond) it would appear as a sphere one foot across.

This bubble illustrates the concept behind the basic units of a granular space-time, although the granules which comprise the space-time in PQR Theory are much, *much* smaller and correspondingly shorter lived. And PQR Theory holds that what we call the space-time continuum is not smooth, as its name implies, but is really composed of a HUGE number of these tiny, short-lived granules. Also, although we can conceive of them as hyperspheres, they don't really have any size or shape or internal structure at all: they're just points with nothing between them, like the points of our two-dimensional grid world. It's only when we try to consider these points as making up our space-time continuum that they seem to have a spatial and temporal extent.

If our space-time is granular instead of continuous, how would we notice this? If the "graininess" was coarse, we might expect to notice a certain roughness, flickering or jumpiness in how things moved, like in an old movie. However, in practice, our sensory apparatus would be subject to the same defects and we would have learned to "smooth out" our perceptions and ignore the jumpiness—just as we do when watching an old movie. In fact, we do this in real life every day, as you can see if you turn your head to look at the seconds display on a digital clock or watch. You will often notice that the clock appears to be stopped for the first second or so before it starts going. This phenomenon (known as saccade chronostasis or the stopped clock illusion) arises because your visual perception experiences a brief time gap as your eyes move to focus on the display. Your brain fills in this time gap with a fixed image of the clock as you first see it, and depending on the precise moment you looked, this can make the first second seem to last longer than it should.

Since our brains are wired to smooth out gaps in our perception, even a fairly coarse granularity of space and time would be hard to notice with our unaided senses in everyday life. And if the texture of space and time was fine enough, then our scientific instruments would not detect the granularity at microscopic scales either. Only when looking at things on the tiniest scales would we start to notice some strange effects. For instance, the electrons surrounding an atomic nucleus would not move around it in a smooth orbit as was once thought, but would appear to jump around from one granule of space-time to another. It would not be possible to predict in advance exactly where an electron would be at any given time, so we would say that the electron's position was indeterminate until it was detected. And this is what happens in real life with quantum physics: fundamental particles don't seem to have definite positions until they're detected.

In fact, at the subatomic quantum level, matter and energy display some bizarre properties. For example, sometimes they appear to behave like particles and at others like waves, and it's not possible to measure all their properties at once. Particles often seem not to have a definite position, and experimental outcomes are not consistently reproducible but vary in accordance with the laws of random probability.

PQR Theory holds that this "quantum weirdness" is due to the granular nature of our space-time, coupled with our insistence on treating it as a smooth continuum.

At the quantum level, the models that agree best with experimental results treat matter and energy as (a) behaving like particles when they interact, but (b) behaving like waves while they move from one interaction to the next. The problem is that we can never detect these waves directly: we can only detect their interactions with our measuring equipment. However, from studying these interactions, quantum physicists have inferred that these waves exist. They have also shown experimentally that they can accurately be modeled by a formula known as Schrödinger's wave equation (after a man who could have invented a boxed cat food called *4½Lives*):

$$i \hbar \frac{\partial \psi}{\partial t} = -\frac{\hbar^2}{2m} \nabla^2 \psi + V \psi$$

This formula describes a wave, but a wave of what? Well, we understand waves in strings, or in water, but Schrödinger waves are *waves of complex probability* in a space-time continuum, which makes them rather hard to understand intuitively. Indeed, while the formulas of quantum mechanics give results that agree very closely with experiments, nobody really understands what they represent. As a result, some quantum physicists have given up even trying to understand their subject. Instead, they simply use the maxim "shut up and calculate."

Do Schrödinger waves exist?

In my view, the physicists' difficulty arises because they have inferred that Schrödinger waves exist, based on their assumption of a space-time continuum. You see, the Schrödinger equation appears to model how matter and energy move against the background of a continuous space-time. But we can also turn this around and say that the Schrödinger equation reflects how *space and time evolve* around quanta of matter and energy. And that begs an even deeper question.

Do space and time really exist?

We have seen how a photon seems to jump from one place and time to another without passing through any of the points in between, and experiments indicate this is also the case with more massive particles of matter. Certainly, we cannot detect a particle without interacting with it; while it is between interactions, we have no way of measuring where it is. So the question becomes, *is it anywhere at all?* In other words, while a particle of matter—or an energy particle like a photon—is in transit between interactions, is it meaningful to talk about the intervening positions in space and time?

You may have imagined our granular space-time as having a bubble at every conceivable point in space <u>and</u> at every conceivable instant of time. In other words, at any given "point" in space there would be a succession of bubbles, one for each "instant" of time. Likewise, at any given instant of time, the whole of space would be filled with these ephemeral bubbles, one at every point. But in fact we don't need all these bubbles, as most of them will represent positions in space-time where and when *nothing happens*—that is, positions which have no interaction between matter/energy particles. We only need bubbles for the positions which correspond to something happening, like a photon interacting with an electron. The other bubbles needn't exist at all for our model to work.

Under this theory, a region of physical space would only exist while something was happening there; correspondingly, time would only appear to pass in a region if something happened there. It's a bit like the old riddle "if a tree falls in the forest and no one is there to hear it, does it make a sound?" But now we're not only asking "if *nothing at all* happens in the forest, does any time pass?" We're also asking "while *nothing at all* happens in the forest, do the trees even exist?" Both questions are essentially unanswerable, since time can only be measured by reference to physical changes, and matter can only be detected when some interaction occurs.

Of course, we're talking about events on *very* small scales of space and time. In a real forest, there are always millions of matter/energy interactions going on every microsecond, and if tiny bits of trees did not exist for minuscule periods of time, we would never notice. (Incidentally, the term *matter/energy interactions* doesn't necessarily imply that we are considering matter as interacting with energy. The term also applies to matter interacting with matter, or energy with energy. At these fundamental scales, matter is the *same thing* as energy.)

If we take the view that time and space are only meaningful at the points where matter and/or energy particles interact, then the Schrödinger equation could be considered as describing the spatial and temporal relationships between these points. This would then allow us to *infer* the existence of space and time—but only at the points where interactions occur. There would be gaps in between these points, which by their very nature would be undetectable. That's not to say that the gaps would all be tiny—for example, photons coming from distant stars could jump across millions of light-years of empty space.

Of course, the space and time would need to be structured in a way that corresponds to our classical notions of space and time, at least at everyday scales. (For example, there should be no "shortcuts" through space, and it should not be possible for anything to travel faster than light or back in time.) Such a model of space and time would in effect be "pinned down" at the positions where matter and energy interact within it. If we wanted to, we could then fill in the gaps between these points by postulating an *imaginary* space-time continuum that would behave in a very similar way to the theoretical model we are all familiar with. The model would tie in with one of the holdings of relativity theory: that there is no *absolute* space or time, but these concepts only exist in relation to a given observer. In the model, space and time only exist in relation to matter/energy interactions, and since the observer is observing, these interactions must be occurring where she is.

So my answer to these questions is that we can say that space and time *do* exist physically, but only at the places and times at which matter/energy particles interact. In between, it's no matter.

3. Results of the Theory

Some principles of PQR Theory

This may be a good point to summarize the key concepts of PQR Theory (including some which I will discuss later on in this book):

- The Universe is a property of the Natural numbers (1, 2, 3 and so on). The arithmetical relationships between these numbers define every event—past, present or future—that has occurred or will occur in the history of the Universe, from its earliest origin throughout all future time.

- Information detailing the entire history of the Universe to date is encoded within the range of numbers from 1 up to some HUGE number. As the Universe ages, this upper limit increases, but at any given time, the Universe holds only a *finite* amount of information.

- Lower numbers are associated with earlier times in the history of the Universe, larger numbers with later times.

- The Universe had a single point of origin, corresponding to the number 1.

- The Natural numbers exist in an abstract mathematical sense; the Universe is their physical manifestation. They go on forever, and so will the Universe.

- The physical structure of the Universe arises from the arithmetical structure of the Natural numbers. Matter, energy, space and time are all interconnected and reflect different aspects of this structure.

- The physical substance of the Universe (matter and energy) exists as tiny packets, each of which reflects a piece of information encoded in the Natural numbers.

- Likewise, space and time exist as individual points which similarly reflect relationships among the Natural numbers; there is nothing between these points.

- Space-time can be mapped in four dimensions, but the resulting model will be *granular*, not continuous. The phrase *space-time continuum* is a misnomer.

- The laws of physics arise from the mathematical relationships among the numbers that define the various elements of matter, energy, space and time, as we will soon see.

Cause and effect

Καὶ ὁ πατὴρ τοῦ τέκνου. (The father is cause of the child.)
—Aristotle (384–322 BC)

Let's now look at what I consider the most basic law of how the Universe operates: the law of *cause and effect*. This glib statement that "effects follow causes" means that the state of the Universe at any particular moment can be influenced by the things that happened at previous times, but not the other way around. Tomorrow's events have no influence on today's situation. This is really a statement about the direction of time—if event A causes event B, then it preceded it in time.

To illustrate: imagine that I drop an egg (cause). Effect: there will then be a broken egg on the floor (at least until the dog eats it). We can turn this around and say that if you see a broken egg on the floor, you can tell that someone must have dropped an egg (and that the dog hasn't found it yet). The point is that everything, in our everyday Universe at least, appears to exist because of what happened to the things that existed before.

This principle applies in PQR Theory too: larger numbers can be derived from smaller ones (for example, by adding and multiplying), which means that later events can be derived from previous ones. The law of cause and effect thus can be considered as inherently arising from the laws of arithmetic.

However, many people consider that the law of cause and effect does not apply at the quantum level, where things appear to happen randomly for no discernible reason. PQR Theory resolves this conflict by saying that *everything* that happens in the Universe *is* the result of prior conditions, but that at the quantum level we can never hope to detect these causes. This is not because our measuring instruments aren't sensitive enough, but because *the very act of measurement* changes the behavior of the system by extracting vital information from it.

For example, we can detect a single photon, but in order to do so, we need a device that absorbs it. We cannot be sure that the photon exists unless we detect it, but once we have detected it, we have used it: its information is gone. Like the photon it detects, our detector is a part of the Universe. It can only detect a photon by interacting with it—so it does not tell us whether or not a photon *exists*; it only tells us that it *captured* one. Scientists have tried to get around this problem by using statistics, running an experiment many times and counting how often a photon is detected. From this, they infer that there is a particular chance of the experiment producing a photon when run, although they cannot predict whether a photon will be detected on any given run. However, unpredictability is not quite the same thing as randomness, as I will discuss later.

PQR Theory holds that although quantum effects in an experiment are not truly random, they *are* unpredictable, because we cannot detect their causes without ruining the experiment. But although these quantum-level causes may not be discernible, they are nonetheless real. The Universe operates deterministically, even though we cannot directly observe the details of its operation. This approach has some interesting philosophical implications, as we will now see.

Free will and determinism

There once was a man who said, "Damn!
It is borne in upon me I am
*   An engine that moves*
*   In predestinate grooves,*
I'm not even a bus, I'm a tram."
—Maurice Evan Hare (1905)

PQR Theory now allows us to reconcile two diametrically opposed philosophical concepts: free will and determinism.

Free will refers to the concept that the self ("you" or "me"—whatever *that* means) has the ability to make its own choices in the present moment. Although our choices may be *influenced* by pre-existing factors such as our physical bodies, our past experiences and our environments, these factors do not *determine* our actions: in the last analysis, we as individuals make our decisions for ourselves.

On the other hand, determinism (or predestination) refers to the concept that everything that happens in the Universe (including the choices its inhabitants make) is ordained in advance. In a deterministic Universe, someone armed with an exact knowledge of its present state and its rules of operation could, in theory, precisely predict its future state at any given time.

Thus, determinism appears to be incompatible with free will. Our thoughts and lives are based on the belief that we have freedom of choice, yet how can we have this freedom if our actions are predestined? No one wants to accept the idea that we are just cogs in some vast machine. Furthermore, our human society requires us to take personal responsibility for our actions, and determinism is in effect a denial of our responsibility. ("Officer, the Universe made me do it.")

Our intuitions and our daily experience strongly suggest that we do indeed have free will. Although someone who knows me closely may often be able to predict my actions, as a being with free will I can always surprise others. For instance, I could go into a sandwich shop and in theory ask for anything that takes my fancy. In practice, of course, my choices are limited, as I will tend to avoid ordering foods I hate or making irrational requests like ordering a helicopter (although I could well ask for a submarine).

Thus, if my regular sandwich maker knows me well, he may be able to predict my order correctly nine times out of ten, but in the last analysis his prediction is really no more than a guess. And this would still be true, even if his shop doorway was fitted with a magic scanner that supplied him with complete information about my mental and bodily state as I entered. Because however predictable I may *seem*, and however well he knows me, I can always surprise him—even if in practice I rarely choose to do so. I can therefore define free will pragmatically as *the capacity to surprise the sandwich maker*. So free will can be viewed as our ability to surprise others (and perhaps, on occasion, even ourselves). Without this ability, we would indeed be mere robots carrying out a predefined program, just like the man who said *"Damn!"* in the 1905 poem above.

Curiously enough, 1905 was also the year of Einstein's famous paper which led to the development of the quantum theory, a theory which has largely supplanted the concept of determinism. Quantum theory says that we can never obtain an exact knowledge of the state of the Universe, and that its rules of operation are not deterministic but *probabilistic*. In other words, events at the quantum level cannot be measured or predicted precisely, but exist only as a range of statistical probabilities. Quantum theory therefore says that our Universe is not deterministic (or at least that we can never know what happens at its most fundamental levels).

PQR Theory on the other hand says that our Universe *is* deterministic—every detail of its workings is precisely specified by the Natural numbers, operating through the medium of one or more formulas. (A quantum physicist would call such a universe *superdeterministic*—which almost sounds precocious.) As explained in the previous chapter, PQR Theory can be harmonized with the well-established results of quantum mechanics by redefining (a euphemism for "abandoning") our old ideas of a space-time continuum. Now I will explain how PQR Theory is consistent with *both* the concepts of free will *and* determinism.

To restate the problem: how can we experience free will in a deterministic universe? How can we be free to choose our actions if they are predetermined for us?

In principle, PQR Theory can provide us with an exact model of the Universe—past, present *and* future—so, with a powerful enough computer, we could in theory calculate the Universe's configuration at any given point in time. Or to simplify the calculations, we could just compute the photons of visible light that impinge on a small area in a brief interval of time. We now have a virtual camera that can take photographs of the past, present and future.

Supposing we program the computer to project this camera a week forward into the Earth's future, and it sends back a picture of me sitting on the beach in Hawaii. That now means I *have* to go to Hawaii and sit on the beach next week, whether I want to or not. Of course, you could argue that the camera would not send back such a picture unless I was already going to go to Hawaii—but that does not solve the problem. Either way, I no longer have the freedom to avoid the beach in Hawaii.

The resolution of this seeming paradox lies in Gödel's theorem, a very subtle and powerful result in mathematical logic. Without getting too technical about the details, this says that any system of logic that is powerful enough to include the rules of arithmetic must either contain a contradiction (a statement that is provably both true *and* false at the same time) or a statement which, although true, cannot be proven using the system. (In practice, there will be lots of these "true but unprovable" statements, not just one.) For example, in mathematics the axiom of choice (see Chapter 2) and all the results that depend on it are known to be unprovable, although they are generally assumed to be true. But in PQR Theory, they are taken to be false. ("Choice is not axiomatic.")

Gödel's theorem tells us that there are some true statements about the Natural numbers that can never be proven mathematically. And since PQR Theory holds that the Universe is a manifestation of the Natural numbers, we can also apply Gödel's theorem to the Universe (a logical system which is powerful enough to include not only the rules of arithmetic but also the mathematicians who formulated them). This tells us that there are true statements about the Universe (facts) that are unprovable *from within it*. (Either that, or the Universe contains contradictory facts, in which case I can prove that you are not reading these words.)

So PQR Theory tells us that we cannot know every true fact about the Universe because we cannot prove every true statement about the Natural numbers. (In fact, mathematicians believe that the vast majority of statements about the Natural numbers cannot be proven true *or* false.)

How does this translate into practical terms? It means that the calculations necessary to predict even the simplest facts about the future will be too voluminous and complex, even for all the world's most powerful computers working together. And this problem can't be solved by building bigger, better, faster computers. Although the future may be fixed, the Universe is simply not big enough to hold a computer powerful enough to predict it. (Quantum effects, combined with the speed of light, provide a practical limit on how small any computer's components can be made and thus how fast it can operate. And don't expect quantum computers to break this barrier—in principle, they will produce multiple answers to any problem, *with absolutely no way of telling which one is correct.* So if your bank had a quantum computer, you'd never be able to obtain your exact balance—just a range of probabilities.)

Thus, although we may know some facts about the future with relative confidence (such as that the sun will rise tomorrow), don't expect to compute which horse will win tomorrow's race. (At least, don't expect results before the race is over. Remember the early computer models of the weather, which took three days to compute a reasonably accurate two-day forecast?) We may one day be able to compute some precise facts about the Universe, but I believe these will always relate to the past, and almost certainly only to the *very* distant past.

Your fate is: to make your own fate

We can now see the fallacy underlying the virtual camera concept above. It assumes the existence of a "powerful enough computer," and

Gödel's theorem implies that we can't build that in our Universe. So although our futures may be fixed, they are *unknowable to us*. In the words of the old song "Que Sera, Sera (Whatever Will Be, Will Be)," the future's not ours to see. But from where we stand today, the future is not preset—we can change it by our actions, and for all practical purposes we can pretend it does not yet exist. Even though our free will may be an illusion, it's a very convenient one. We may be mere cogs in this vast machine, but we are cogs with a part to play.

So viewed from within, the Universe *appears* to grant us free will. On the other hand, if we could view it from outside, it would appear deterministic. But of course we can't do this, because there's no way to get there, no place to stand, and nothing to see anyway (no photons go there). The bottom line: within the Universe, our free will is real. It only becomes an illusion when we try to consider the Universe *from the outside*. That is the ultimate paradox of reality.

Are you for real?

All this might just be an elaborate simulation, running inside a little device sitting on someone's table.
—Jean-Luc Picard (Stardate 46424)

If you play video games, you'll be familiar with the idea that computers can be used to create artificial worlds. As computing power has increased over the last twenty-five years, these virtual realities have become more and more detailed and realistic. This idea was taken to its logical conclusion in the movie *The Matrix,* in which the whole world was revealed to be a vast computer simulation, created by machines two hundred years in the future. The entire human race was trapped in this collective illusion, and people went about their daily business in blissful ignorance that they were, in reality, the slaves of the machines.

Could we all be "Sims," living inside a computer simulation? And if we were, would we realize it—or would we have been programmed not to? What would it be like to live inside a simulated reality?

Well, computers simulate reality by taking a grid of points in space and computing what would be there at a given time, and then projecting this forward from one moment of time to the next. (Typically this is done by tracking the grid position of each object in the simulation at each moment of simulated time.) But being finite machines, computers can only deal with a finite number of grid points and a finite (rather than infinitesimal) interval between successive moments of time. In a very powerful computer, the grid may be very finely grained and the time intervals very small, but they can never be infinitesimal. So if we were living inside a simulation, we would notice that space and time are not infinitely divisible. The limitations of the computer's information-processing power would limit our ability to observe very small quantities of space and time.

As an analogy, consider a digitally recorded movie. The recording consists of a large (but finite) amount of information representing events over a finite span of space and time. The events in the three-dimensional space are represented on a two-dimensional screen; the image is divided into small pixels (the screen contains millions of these); each pixel is divided into three primary colors; each color's brightness is mapped on a scale from 0 to some large number; and the time is divided into short intervals. In this way, the whole sequence of events can be transformed into a series of binary digits, which in turn are recorded on a suitable medium such as a DVD. With the proper device sitting on our table, we can play back the DVD and watch the movie. Although the actual information on the DVD is merely a series of ones and zeros, we have the illusion of watching a slice of real life. Captain Picard did not know how truly he spoke (although presumably Patrick Stewart did).

Note, however, that the information in the movie is *quantized* by the process outlined above. If there are thirty frames of information per second, we can know what happens and where everything is in each frame, but we cannot see what happens in between the frames. An event lasting less than 1/30 of a second (such as a flash of light) may be captured by the camera or it could in principle "fall between the cracks" and not be recorded at all in the movie. And moving objects tend to blur so that we can't tell their precise positions in the frame. Likewise, each frame of the movie is pixelated, so we can never know the *exact* position of an object, even one that's not moving—we can only measure it to the nearest pixel. Nor can we reliably detect objects that would be smaller than one pixel in size on the screen. Very small objects (and larger objects a long way off) may not show up at all, not even as a single pixel.

Following this analogy, we would expect that if we were living in a computer simulation, we would find uncertainty in our measurements, especially when looking at small objects over short periods of time. And indeed, the real world turns out to be very much like this. Quantum theory teaches us that we cannot accurately know both a particle's position and its velocity. The more accurately we measure one, the less accurately we can determine the other. And to reiterate: this is *not* due to inadequacies in our measuring equipment, but is a fundamental constraint arising from the fact that the particle contains or comprises only a finite quantity of information.

So at first blush, it would appear that we could indeed be living inside a computer simulation.

What sort of simulation might this be? Well, one kind of model that has been extensively studied is the *cellular automaton*, a well-known example being Conway's Game of Life. This consists of a two-dimensional grid of cells, each of which can be either "on" or "off" at any given time. Time is not continuous in this game but is measured in discrete clock ticks. The game has a rule which specifies, based on the status of a cell and its neighbors at any given time, whether the cell will be "on" or "off" at the next clock tick.

The surprising thing about the Game of Life is that its single simple rule, applied over and over again to the grid, can produce patterns that behave in amazingly complex ways. (This is the central theme of *A New Kind of Science*, Stephen Wolfram's exhaustive study of cellular automata, a massive tome that tells you far more than you ever wanted to know about the subject.)

So if simple rules can produce complex behavior, could the Universe be a cellular automaton? I would answer *no*, for the following reasons:

First, cellular automata utilize a regular grid, which will have "preferred directions" in which phenomena such as light rays would propagate more easily than in others. No such preferred directions have been observed in reality: space appears to be *isotropic*, meaning that it looks the same in whichever direction we turn. If there were a grid, objects would appear different when looked at from different directions (e.g., along the grid lines vs. at an angle to them).

Second, such a grid would behave like the "luminiferous aether" we'll be discussing soon: the speed of light would be different when measured along or across the direction of Earth's movement through the grid. However, the famous Michelson–Morley experiment showed this not to be the case, thus discrediting the aether theory. So it would appear that the Universe cannot be a cellular automaton, nor (for the same reasons) could it be any other kind of computer simulation that uses a grid.

But could it be some other type of computer simulation that avoids reliance on a grid? Once again, I would answer *no*, but this time for a more subtle and powerful reason: a simulation of the Universe would require a giant computer in order to operate. Where would such a computer exist, and what would it be made of? It could not exist in our Universe or it would have to *simulate its own operation* in real time—a logical impossibility. Such a computer could only exist in another universe, even bigger than ours. And what would *that* universe be made of? This approach does not solve our philosophical problems, it simply makes them bigger. (And it's no answer to say that it's terminals all the way up.) So, in the spirit of Occam's razor, which I turn to next, I say we must abandon this idea and thankfully reach the conclusion that we are not Sims. Even if we are in the last analysis creatures composed of information, that information is real and natural, not artificial and simulated.

Shaving Schrödinger's cat with Occam's razor

The medieval philosopher William of Ockham (c.1288–1348) is associated with the principle known as Occam's razor, which says that "entities should not be multiplied needlessly." In other words, theorists should follow the advice that "everything should be made as simple as possible, but not simpler." (A saying attributed to Einstein, along with "I never said half the crap people said I did.")

Old William's principle is still applicable to modern physics. I am naturally suspicious of any theory that postulates the existence of mysterious influences at work behind the scenes, especially ones which cannot be verified experimentally. The history of science is littered with discarded ideas like phlogiston, caloric and the luminiferous aether that were postulated to explain phenomena like fire, heat and light waves. These hypothetical substances could not be directly detected and were ultimately shown not to exist, whereupon new theories were devised that made them unnecessary. There was also the mythical planet Vulcan, originally designed to explain an anomaly in Mercury's orbit but now totally exploded (or, fictionally, *imploded,* so that now it only explains Spock's pointy ears).

The moral to be drawn is this: if you can't detect it, it probably doesn't exist, and you should look for a new explanation. Any theory which entails layer upon layer of complex undetectable processes is doubly suspect. So let us take Occam's razor in hand and apply it to present-day physics. Which of today's cutting-edge ideas will end up on the cutting-room floor of history, discarded and forgotten like the aether? Remember, the most recent theories are not necessarily the best, they are simply the ones that haven't yet been around long enough to be discredited.

Quantum mechanics

Plenty of descriptions and explanations of the quantum theory are available elsewhere, so I will not waste too much time repeating them here. I will simply say that quantum theory has been very good at describing *how* things work at the subatomic level. Its predictions have been verified experimentally to a remarkable degree of accuracy. But it has totally failed to explain *why* things work the way they do. At the subatomic level, quantum weirdness takes over: while we can accurately *model* the behavior of photons, electrons and the like, we are totally unable to *explain* what is going on. It is as if the uncertainty principle governs the theory of quantum mechanics itself—the more precisely we can predict something, the less we can understand it.

The basic mathematics of quantum theory can be formulated in a number of ways. For example, Schrödinger's wave equation assigns a Complex number to every point in space-time and then relates how these numbers vary from place to place and from time to time. An alternative formulation involves analyzing the transformations of infinite-dimensional matrices. But whichever way we try to model the behavior of even a single particle, we find that we need to mathematically analyze an infinite range of numbers. Talk about multiplication of entities! And yet when we try to *measure* the particle, we find (by the uncertainty principle) that we can only extract a very limited amount of information from it. Were all those numbers really necessary, or should we be looking for a radically simpler formulation?

Interpretations of quantum mechanics

The calculations of quantum mechanics may be complex, but they work well. The real trouble begins when we try to interpret what is going on. According to the standard "Copenhagen" interpretation, the Universe operates randomly at its most fundamental level: each particle exists as a wave function that defines the probability of its being detected in any particular location. We cannot detect this wave function directly, since any act of measurement "collapses" it down to a single value. Thus, a particle has no definite existence but exists only as a range of possibilities, a probability wave smeared out in space, until it is observed. The act of observation destroys the wave. This interpretation leads to such philosophical absurdities as Schrödinger's cat, which is shut in a box along with an apparatus rigged to kill it when a radioactive isotope decays—a quantum-driven random process with a 50% chance of occurring within an hour. According to this interpretation, the cat manages to be *both dead and alive* at one and the same time, until someone opens the box after an hour to check on it. It is only when the cat is observed (thus collapsing its wave functions into a definite state) that the Universe "decides" whether it is alive or dead.

To my mind, this is hubris of the highest degree. To say that a system has no objective reality until it is observed is like saying that a tree falling in the forest only makes a noise when someone is near enough to hear it. Why should the laws of physics only apply in the presence of a human observer? And to say that two possible cats, one dead and one alive, occupy the box for an hour is to multiply entities needlessly. Old William would *not* have approved. The cat is a live cat unless and until the isotope decays, after which it is a dead cat. We shouldn't say that both the live cat and the dead cat are in the box just because we don't know which one it contains. (And in any case, while it was alive *the cat* would know it was in the box.) To be fair to Schrödinger, I should mention that he invented his cat in order to *discredit* the Copenhagen interpretation.

Of course, PQR Theory holds that our Universe is deterministic, so there can be no *truly* random events, only unpredictable ones. But even setting aside this philosophical distinction, there is still a real difference between the concepts of randomness and unpredictability as they apply in everyday life. While all random events are indeed unpredictable, not all unpredictable events are random. For example, lottery numbers are as random as human ingenuity can make them, so that they are unpredictable (or the lotteries would very rapidly go out of business). On the other hand, stock prices and ballgame scores may be so unpredictable as to *appear* random, but we know them not to be, because we realize how they are determined by human actions.

Our brains are wired to look for patterns in everything we see, since recognizing a pattern helps us to predict how it will develop and to respond appropriately. (From an evolutionary perspective, the ability to quickly distinguish a live saber-toothed tiger from a dead one would have been a useful survival skill.) As a result, if we see a process where we don't recognize a pattern and can't discern its causes, it's natural for us to consider it random. But is this because no pattern exists, or have we just not found it yet? Consider the sequence of digits 2, 6, 4, 5, 7, 5, 1, 3, 1, 1, 0, 6, 4, 5, 9, ... It may *look* random, but once we recognize it as the square root of 7 we know how to predict it—and, of course, that it isn't random.

This highlights the difference between randomness and unpredictability: unpredictable processes follow no *discernible* pattern, but truly random ones follow *no pattern at all*. And this is where I feel the Copenhagen interpretation is flawed. It confuses these two concepts.

To illustrate this in everyday terms: I flip a coin so it lands behind a screen. Later, we look behind the screen to see if it landed heads or tails. Before flipping the coin, we can say there is a 50/50 random chance of getting either outcome. But once the coin lands, the outcome of the experiment is no longer random but *determined*—even if it remains unpredictable until we look at the coin. The confusion arises when we say, before looking at the coin, "there's a 50/50 chance that it *came* down heads or tails." This is misleading: the coin has already landed, so it's either heads or tails. Chance no longer has anything to do with it. Yes, there is a 50/50 chance that *when we look* at the coin *we will see* heads or tails; but the appearance of chance lies solely in the fact that we don't yet know how it landed. And this discussion concerning flipped coins applies equally to entangled photons.

This allows us to resolve the famous Einstein–Podolsky–Rosen paradox, which says (in its simplest terms) that if the result of measuring a photon is truly random, then information can appear to be transmitted faster than light. The experiment (which can also be done using elec-trons) involves sending two randomly polarized photons off in opposite directions to two distant observers, whom we will call Carol and Ted (because Alice and Bob are fed up with doing it). The equipment can be arranged so that the two photons are created and sent out in an "entangled" state, meaning that if Carol and Ted perform the same measurement on their photons, they are (in principle) guaranteed to get opposite results. The problem arises because Carol and Ted both believe their results to be truly random, but once Carol performs her measurement, Ted's result ceases to be random—even though no message could have traveled from Carol to Ted in the time available. Einstein found this hard to accept, calling it "spooky action at a distance."

In PQR Theory, this seeming paradox is resolved by saying that, at the instant the photons leave the polarizer, each one's polarization state is precisely determined. (A quantum physicist would call this a global hidden variable.) And this polarization precisely determines what results Carol and Ted will obtain when they perform their measurements. Since their results are unpredictable, they consider them to be random, even though that randomness disappeared at the instant the particles were sent. (Of course, from the photons' viewpoint, that would also be the instant they arrived; but this analysis is also applicable to slower-than-light particles.)

The point is that applying probabilities to past events can be misleading. What are the odds on yesterday's race? They're 100% in favor of the winner, even if we don't know which particular horse won.

To put this in quantum-mechanical terms, the Schrödinger wave equation does not describe the underlying processes at work but merely sets limits on how much we can know about them. (So it is the passage of time, rather than the actual act of observation, which would cause the collapse of the quantum waveform.) And while probabilistic interpretations of quantum mechanics are useful for considering the possible outcomes of experiments we have yet to perform, they are inapplicable to past events—unless they give a postdiction which is strongly at variance with the known outcome, in which case we can safely say there is something wrong with our model.

The perverse multiverse

Another possible view of quantum mechanics is the so-called many-worlds interpretation. This rejects the idea that the waveform collapses when we observe Schrödinger's cat. Instead, *the entire Universe, including the observer,* splits into two: one possible Universe in which the observer sees the cat as alive, and a parallel Universe in which her counterpart sees the cat as dead. And this splitting process does not only happen when we put a cat into a box, but it goes on all the time, in every part of every conceivable universe. As a result, there are a myriad of possible universes, all of which are forever splitting into more and more alternate realities. Our Universe follows just one of the impossibly many branching paths in this hyper-universe; all other paths are undetectable to us. (A cosmological variant of this theory hypothesizes that our Universe is just one of many which were formed in the Big Bang but which later split apart and went their separate ways.) The mere thought of this "multiverse" would have had old William cutting his throat.

Quantum field theory and electrodynamics

And when we start looking at quantum field theory and quantum electrodynamics, which model the interactions between different particles, things get really weird. According to these theories, we cannot know which route a photon or an electron takes as it moves from place to place. Instead, we must calculate and combine the probabilities for *every possible route* the particle could take. And then we must allow for the possibility that the particle does some weird things along the way, such as emitting and then reabsorbing one or more "virtual" particles. These are particles so short lived that they cannot be directly detected, they can only be inferred from the theory. (Notice a pattern here?) We must also allow for the possibility of virtual particle–antiparticle pairs being spontaneously created and then annihilated. And after doing all these calculations, we must take one more step that, from a mathematical viewpoint, is *totally invalid.* Known as renormalization, this involves canceling out the infinities which are thrown up by the calculations. And yet, it gives us answers that agree fantastically well with experiments.

What is going on here? Why do we need to assume the existence of all these undetectable entities in order to reach the "right" answer? Why can't physicists interpret their own theories? It's all enough to make old William turn in his grave (which presumably contains not only the dead Occam but also the buried-alive one).

Dark matter

Occam's razor can also be used outside the submicroscopic quantum realm. We can apply it to astronomy, where it is speculated that only about 4% of the matter and energy in the Universe is detectable. The remaining 96% is said to consist of "dark" matter and energy.

Cosmologists make this arbitrary assumption in order to get their theories to agree with their observations. Now, which do you think is more likely: (1) that the Universe consists mostly of some mysterious undetectable forms of matter and energy, or (2) that the formulas used in cosmology are somehow off by a factor of about 24? I think I know what old William would have said.

Quarks

On a related note, physicists believe that the atomic nucleus is comprised not of protons and neutrons but of smaller particles called quarks, which are joined together by other particles called gluons. Again, this model agrees very well with experimental results, but it suffers from a problem: nobody has ever detected a single quark. They only seem to be detectable in twos and threes (and perhaps also fives), which raises the question: *do they really exist as independent entities?* Gluons are even more slippery, as they seem to exist only as virtual particles. Maybe the model needs to be reinterpreted. For example, a piece of paper contains two sides and a triangle has three corners, but are the sides and the corners really separate objects? Or do they only exist as part of a larger whole? Maybe particles like quarks are merely facets of a larger underlying reality. I will leave this as an open question for now.

Applying Occam's razor to PQR Theory

Having read the foregoing paragraphs, you may be asking yourself "why should I accept PQR Theory?" Surely postulating an inherently undetectable matrix of granular space and time is just as bad as all the other questionable ideas mentioned above? What would old Occam have thought?

The answer is in two parts. First, we only need to assume the existence of the Natural numbers in order to build the models of PQR Theory. Providing we can find a suitable formulation for a granular structure of space and time within the Natural numbers, we're not really inventing anything new, just using what already exists.

Of course you might argue that the Natural numbers don't have a *physical* existence—you can have seven apples, but can you have the seven without the apples? To this I would answer that the Universe itself is the physical embodiment of the Natural numbers. I realize there's a certain circularity to this argument, but that may be inherent in the nature of creation itself: the Universe may only exist *because it can*, and not because of any external causative event.

And second, Occam only used his razor to cut off his beard, not his head (yes, he *was* clean shaven). He said we shouldn't complicate things *needlessly*, not that we should cut out *everything*. The goal of PQR Theory is to underpin our current formulations of relativity and quantum theory (which are based on a space-time continuum) with a single, simpler underlying formulation based on Pythagorean number-theoretic principles. Eventually, the space-time continuum would be seen as a needless entity like the aether, useful perhaps for mathematical calculations but only an approximation to the underlying granular reality. (Incidentally, this view could solve the renormalization problem mentioned above, by replacing the infinities with very large—but nevertheless finite—numbers.) I wonder: which explanation would Occam then choose?

As Einstein wrote in 1920 (yes, he really did):

Es ist das schönste Los einer physikalischen Theorie, wenn sie selbst zur Aufstellung einer umfassenden Theorie den Weg weist, in welcher sie als Grenzfall weiterlebt. (No fairer destiny could be allotted to any physical theory, than that it should of itself point out the way to the introduction of a more comprehensive theory, in which it lives on as a limiting case.)

4. Some Unanswered Questions

The greatest fool may ask more than the wisest man can answer.
—Charles Caleb Colton (1820)

Before starting to look at model universes, I want to consider some cosmological questions whose answers are typically taken for granted. I'm not saying that accepted wisdom is wrong in every case, just that we should be careful not to accept it blindly. So consider these more as talking points than as serious criticisms of existing theories.

Are the constants of physics changing? And if not, why not?

Are the so-called universal constants of physics really as universal and unchanging as physicists assume?

The scientific method relies on the principle that experiments are reproducible. In other words, the same experiments, performed by different people at different times and different places, should produce the same results. Any variations should lie within the experimental margin of error.

This principle has been widely confirmed to hold good in practice, and as a result, physicists have been able to measure a large number of physical quantities with great precision and confidence. Many of these quantities, such as c, the speed of light, h, Planck's constant, and e, the charge of an electron, appear to be always the same regardless of who does the measuring, or when, or where. Accordingly, physicists call these quantities "constants" and base their theories on the *assumption* that these numbers never change.

For example, the speed of light is 299,792,458 meters per second, Planck's constant is about 6.626068×10^{-34} joule-second (pretty darn small on any scale), and the electron charge is about 0.16021765 of an attocoulomb. (Not to be confused with the electric charge, which is about 17 cents per kilowatt-hour at current prices, or over $100 at battery prices.)

Why these particular numbers? Well, of course they depend on the system of units we have chosen for our measurements. In many cases these units were chosen more or less arbitrarily at some point in the past.

For example, the second was originally derived as 1/86,400[th] of the time the Sun apparently took to go round the Earth, and the meter was calculated as 1/10,000,000[th] of the length of a line from the Equator to the North Pole via Paris. Nowadays, we find it more precise to define a second as the amount of time taken for a certain resonance of the caesium-133 atom to occur 9,192,631,770 times, and a meter as 1/299,792,458[th] of the distance traveled by light during this time. That way, we can state the speed of light *exactly,* and we can also be sure that it is not changing: if anything is changing, it is the length of the meter.

Not surprisingly, most physical constants turn out to be somewhat awkward numbers when we express them in terms of these rather arbitrary units. However, some physical constants do not depend on our choice of units. These are the so-called dimensionless constants: such things as the ratio between the mass of a proton and that of an electron (about 1836.152672) and the "fine structure constant" which is about 1/137.035999. These numbers apparently represent inherent properties of the Universe around us, *but nobody knows why they have the values they do.* Nobody has been able to find a plausible mathematical formula governing these values, which raises a question in my mind: *could they be functions of the age of the Universe?*

If so, they would be changing, albeit probably very slowly, as the Universe ages. Can we be sure that we've been calculating these values for long enough, and with enough accuracy, to have detected any possible change? And if they are changing, how would that affect our extrapolations of the history of the Universe?

There is still one unit of measurement that does not yet have a precise physical definition, and that is the kilogram. Although this may soon change, this is still defined as the weight of a particular metal cylinder that has been kept in a vault on the outskirts of Paris since 1879. In order that this standard can be used worldwide, many copies of this cylinder were made in the 1880s and distributed to various nations around the world. Periodically, they are brought back to Paris and checked against the original, but after a hundred years, a strange phenomenon has become apparent: *the weights of these "standard" kilograms are gradually diverging.* The divergence is tiny—of the order of only 50 parts in a billion over a century—but it is real, and physicists have been unable to come up with a satisfactory explanation.

One possible theory is that the cylinders were manufactured from a

90% platinum–10% iridium alloy. Now platinum has an atomic weight of 195.084 but iridium's is only 192.217—about 1.5% lower. But when the cylinders were manufactured, the 90%–10% platinum–iridium ratio was only controlled to within a tolerance of ± ¼%. Could the atomic weights of platinum and iridium have changed by slightly differing amounts over the last hundred years? If so, then the platinum-rich cylinders would naturally exhibit a divergence in weight from the iridium-rich ones. However, this explanation is unlikely to account for the entire discrepancy, so there is probably some other influence at work.

When we talk about "universal" constants, we should bear in mind that all of our experiments have been performed in a *very* small region of the Universe. If some of the "constants" were different elsewhere, would we have a way of telling? Yes, the Universe *looks* the same in every direction, but how can we be sure that we'd find no differences at all if we could go a long way off?

How old is the Universe?

The age of the Universe is now estimated by cosmologists to be about 13.800000002 billion years. (Well actually, a couple of years ago they estimated it to be about 13.8 billion.) This illustrates the point that from one year to the next, the proportionate change in apparent age is going to be minuscule. Very few of the constants of physics have been measured to this degree of accuracy. If they are in fact changing, gradually increasing or decreasing as the Universe ages, would we yet know this? For example, a parameter that varies in proportion to the age of the Universe will only have changed by about 0.0000007% over the last hundred years. Such a small variance, even if detected, would almost certainly be attributed to experimental error. So can we be confident that the constants of physics really are constant?

All the calculations performed by cosmologists are based on a tacit assumption that their constants are unvarying, but if they are wrong, today's physical processes may run at vastly different rates than they did in eons gone by. In that case the cosmologists' estimates are going to be off, perhaps by a large factor. So we should view the stated age of the Universe as an *apparent* age, based on the unsupported assumptions that these constants are really constant.

I do not for a moment suggest that the Universe was created in the year 4004 BC, or as recently as last Monday. But if the "constants" of physics are changing, then it could have been a fair bit less (or more) than 13.8 billion years ago. Even the concept of a year becomes somewhat elusive when you are looking back to a time long before the Earth started going round the Sun. So we should treat these estimates with caution.

Timeline of the Big Bang

Our whole universe was in a hot dense state,
Then nearly 14 billion years ago expansion started...
—Barenaked Ladies (2007)

Cosmologists spend hours discussing what happened in the first millionth of a second or so after the Universe came into existence. But how meaningful is this sort of a timescale, so early in the Universe's history? After all, there were no clocks back then—not even any caesium atoms to count the vibrations of. These early-Universe timescales must therefore be regarded as no more than an extrapolation of how fast things happened, based on our present-day physical constants. The underlying processes driving the Universe may have operated a lot slower (in some sense) than they appear to do today. If we accept that as a possibility, then the Big Bang may not have been a sudden rapid expansion at all, but a much steadier and smoother "Big Blossoming"—a blossoming that is still going on today.

Is energy conserved over long time periods?
One of the fundamental laws of science is the principle of conservation of energy. This states that within any closed system, the total energy (including the energy content of matter) cannot be changed by anything happening within the system. The energy may change its form (for example, potential energy becomes kinetic energy when a ball is dropped), but the total amount will always remain unchanged. This principle has been widely verified by numerous experiments—but does it hold good over long periods of time? What if the total energy in the Universe has not been the same for all time, but is increasing in line with its age? Would we notice this? As mentioned above, even over the course of a hundred-year experiment there would only be a tiny percentage variation in energy, so in all probability the change would not be detected.

So if energy is conserved locally but still grows gradually over long periods of time, we would not know this. Measurements made over a short time period (that is, short in comparison to the age of the Universe) would *appear* to demonstrate conservation of energy, but in fact this would only be an approximation.

If energy has not been conserved since the start of time, this also would undermine the concept of the Big Bang. Maybe all the energy in today's Universe did not come into existence in a single instant, but arose as the result of a much more gradual process, a process that is still ongoing.

Is space expanding, or is it subdividing?

We are told that the Universe is getting bigger: the distant galaxies are moving away from us. Another way to look at this would be to say that the Universe remains a constant size, but the space within it is *subdividing*. These two approaches are in fact mathematically equivalent; all we need to do to convert from one to another is to apply a scaling factor that increases over time.

To illustrate with a simple example: imagine a small model universe which only contains 1,000 points or granules of space at a certain time. As time passes, new granules are formed in between the existing ones, so that sometime later, there are 2,000 granules of space. We can look at this in one of two ways: we can either say that the granules have remained the same size and the total volume of space in the model has doubled, *or* we can say that the total volume has remained unchanged but the granules have all halved in size.

Since the physical processes in such a universe would be driven by the granularity of its space and time, its inhabitants might recognize that they and the things around them were getting smaller as the space in their universe became more finely grained. However, they would probably find it more convenient to conduct their science on the basis that the fabric of their space-time was unchanging. They would then consider themselves to be living in an expanding universe, just as we do.

When discussing the model universes in Part II of this book, it will sometimes be convenient to use one approach and sometimes another, but bear in mind that they are equivalent.

Some unanswerable questions

I keep six honest serving-men
  (They taught me all I knew);
Their names are What and Why and When
  And How and Where and Who.
—Rudyard Kipling (1902)

Some questions about the Universe cannot be answered satisfactorily, because they are framed in terms that are ultimately meaningless when viewed through the lens of PQR Theory. So the answers I give may be flip, but they're the best I can do—and hopefully they point out the difficulty behind these questions.

What is the Universe? Everything and nothing. Everything, in the sense that it comprises all matter and energy. Nothing, in the sense that its essence is an abstract concept, the concept of number.

Where is the Universe? Everywhere and nowhere. Everywhere, in the sense that every place is a part of the Universe. Nowhere, in the sense that it is not located *in* any place.

Why does the Universe exist? Because it contains philosophers who think it exists, and they can't possibly be wrong. (The philosophers, in turn, exist because they think.) Alternatively, the Universe exists in order to permit philosophers to ask these questions. Either way, we're here because we're here because we're here...

When did it come into existence? Never. There was no time before the Universe existed.

What was there before the Big Bang? *Nothing.* This is the hardest concept to accept. There was no empty void in which the Universe would form: there was simply *no space or time.* PQR Theory holds that space and time only exist in the presence of matter and energy. And when people say "but there must have been *something* there then," I can only answer "there *was* no 'there' there and there *was* no 'then' then."

Does God exist?

Si Dieu n'existait pas, il faudrait l'inventer.
(If God did not exist, it would have been necessary to invent him.)
—Voltaire (1694–1778)

The Creation of Adam—but who is creating whom?

I've saved the biggest question for last. Unfortunately (or fortunately) it is beyond the scope of PQR Theory to answer.

In the previous chapter, I discussed how Gödel's incompleteness theorem states that any consistent system of logic contains true statements that cannot be proven from within the system. (We are only discussing systems that are powerful enough to generate the Natural numbers.) There is an extension to this theorem (sometimes known as Gödel's second incompleteness theorem) which states that you cannot prove the consistency of a system of logic from within it. (Consistency means that no statement in the system can be proven both true and false.) In other words, you cannot use the system to prove its own consistency (unless, paradoxically, the system is *inconsistent*, in which case you can use it to prove anything).

What does this tell us about the Universe as a logical system? Well, first of all let's assume that the Natural numbers are consistent, that there are no contradictions in arithmetic. (I'm not even going to *consider* the alternative here—that would be another story altogether.) Then Gödel's second theorem tells us that the Universe cannot contain a proof of its own consistency. Thus, although we may *believe* that truth cannot contradict itself, this is something which we, as creatures of this Universe, can never prove: we must take it on faith. (So when Pontius Pilate asked "what is truth?", he was asking the unanswerable.)

Can miracles occur? If we define a miracle as an occurrence that cannot be explained by the laws of nature, this theorem tells us that we can never prove miracles to be impossible. And of course PQR Theory can never show that miracles *do* occur (or they would not be miraculous). So it is up to us to choose what to believe. Any *apparent* miracles can be attributed either to divine intervention or to our own imperfect understanding of how the Universe operates, depending on our beliefs. (And our beliefs change: at one time, eclipses were considered miraculous. We know better now.)

The same principle applies when we consider the miracle of creation, in other words, the idea of God as creator of the Universe. God, as framer of the laws of nature, cannot be explained by those laws, or we would be faced with a chicken-and-egg situation. Nor can the existence of a creator be disproved by PQR Theory, which only talks about what happens within the physical Universe. Even if the theory could explain every phenomenon ever observed in the Universe, this still would not *prove* that there was no other Force at work behind the scenes. In short, PQR Theory shows that *we can neither prove nor disprove the existence of God*. Ultimately, this must be a question of faith, of what we choose to believe.

So what follows is my own personal take on the subject.

As a species, we choose to believe what suits us. Any race of humans who believe they can fly off cliffs is likely to face rapid extinction. So we believe we cannot do this, even though most of us have not tried it for ourselves. Likewise, when it comes to belief in God, the god-fearing nations have had an evolutionary advantage over the godless. Even though many people nowadays question or deny the existence of God, historically our society has been founded on a belief in a creator who has given us laws by which to conduct our lives. We are encouraged to obey these laws by the promise (explicit or implied) of divine reward, with the accompanying threat of divine retribution if we disobey. Our civilizations have flourished enormously under the social order and discipline that flow from such beliefs. And faith and discipline operate to strengthen a nation's forces in times of war. So it should come as no surprise that successful civilizations are founded on faith, and that many of us, as the products of these civilizations, do have a belief in God.

But why does any individual believe in God? A cynic would say that one's belief in God is the product of evolution, and some might even go so far as to say that *the whole idea of God* is the product of evolution. No doubt many people do believe in a deity simply because they have been taught that this is the "right" thing to do. A pragmatist would say he chooses to believe in God because this is the less risky option: by believing in a nonexistent god we lose nothing, but by ignoring a God who exists we run the risk of losing our eternal reward (and perhaps also incurring eternal perdition). Likewise, many people who don't consider themselves superstitious nevertheless choose to believe in God on the basis that it costs nothing and might improve their karma in life. Some idealists may feel a duty to believe in God in order to participate in making the world a better place. And many of us would argue that we believe in God in order to fulfill our inborn spiritual needs.

All these viewpoints have some validity. My own personal belief is that the Universe is too wonderful a place to have come into existence as the result of blind chance; but in the last analysis, this is really only a gut feeling—and I can hardly expect anyone to change their beliefs based on my gut feelings. You will just have to decide for yourself.

PART II:
HOW ONE CAN CREATE A UNIVERSE

שְׁמַ ע, יִשְׂרָאֵל: יְהוָה אֱלֹהֵינוּ, יְהוָה אֶחָד

(Hear, O Israel: The Lord our God, the Lord is One.)

—Moses

5. Model Universes

Having laid the philosophical foundations of PQR Theory, it's time to see how they can be put into practice. The following chapters describe how a series of models can be built up using arithmetic, geometry and algebra; how these models contain elements which resemble time and space, energy and matter; and how these elements behave according to laws similar to those that govern the physical Universe in which we all live.

These models are not only provided to illustrate and explain the principles I have been discussing. They are also an attempt to validate the ideas underlying PQR Theory by showing how its concepts can be used to build workable model universes.

Up to this point, I've been very gentle in my use of mathematical terminology and formulas, but now it's necessary for the gloves to come off so I can show exactly how these models are built. I hope you will find it possible to understand the descriptions of the models, even if you don't care to examine the formulas too closely. But I do feel that the mathematical details have to be included in order to give credibility to the models. Fortunately the math is not too advanced, even though the models make use of some concepts that will not be familiar to the reader. Some of these concepts are from branches of mathematics that are not widely studied; others were developed specially to build the models. But please feel free to skip over the mathematical details if you don't find them interesting—I won't be offended.

I'm not claiming that these models necessarily provide an accurate representation of our Universe. However, I do believe that models along these lines can be used to build universes similar to ours, and that they illustrate the *principles* by which such universes can be constructed from numbers. The *details* of how our own Universe is constructed may turn out to be somewhat different from the details of the models described in the following pages, but I believe that there are enough parallels between the behavior of these models and the behavior of our own Universe to show that the same principles apply throughout. So while these models' structures can be analyzed and their properties known with mathematical precision, remember that their applicability to our physical Universe is a matter of speculation—at least, until it can be verified experimentally.

We will start with a simple model which has only one dimension of space and evolves in a way that corresponds to the passage of time. This model is L1, the Level One model. We follow this up with a second-level model, L2, which has two dimensions of space and one dimension of time. Then we examine a third-level model, L3, with three dimensions of space as well as the dimension of time.

Each level builds on the one before it: the space in the L1 model is a simple line segment joining two points; the space in the L2 model is an equilateral triangle whose edges are three copies of the L1 model; and the space in the L3 model is a regular tetrahedron (a solid shaped like a pyramid, but with a triangular base instead of a square one) whose faces are four copies of the L2 model.

Thus, each level model contains copies of the lower-level models: in fact, they form its edges. This is very similar to the idea of a "holographic universe" which some theoretical physicists have speculated on, and also similar in a way to Plato's cave, as the objects contained in each level in some sense resemble shadows of higher-dimensional objects contained in higher levels.

The L1, L2 and L3 models can be built up in a number of different ways, but these ways are equivalent, since they all produce the same structure at each level. One way is a geometrical approach that uses circles, spheres or hyperspheres that touch each other. Another way is purely by considering groups of numbers—the arithmetical approach. A third approach is algebraic—by looking for whole-number solutions to an equation and finding how these are connected. (These connections can also be charted as a mathematical graph, giving us yet a fourth way of examining the models' structures.)

The following table shows the different ways of constructing the models that are explained in the next three chapters.

A comparison of the models		
L1 (Level One) "Fareyland"	**L2 (Level Two)** "The Flatrack"	**L3 (Level Three)** "The Four-D Froth"
1 dimension of space + 1 dimension of time	2 dimensions of space + 1 dimension of time	3 dimensions of space + 1 dimension of time
Geometrical Formulation:		
Ford circles tangent to line segment; 2-way symmetry	Ford spheres tangent to equilateral triangle; 6-way symmetry	Ford hyperspheres tangent to regular tetrahedron; 24-way symmetry
Arithmetical Formulation:		
Duads: **a:b>y**	Triads: **a:b:c>y**	Tetrads: **a:b:c:d>y**
Algebraic (Diophantine) Formulation:		
Cartesian Trios: $(w + x + y)^2 = 2(w^2 + x^2 + y^2)$	Soddy Quartets: $(v + w + x + y)^2 = 3(v^2 + w^2 + x^2 + y^2)$	Gosset Quintets: $(u + v + w + x + y)^2 = 4(u^2 + v^2 + w^2 + x^2 + y^2)$

In the pages that follow I describe the L1, L2 and L3 models. (Theoretically one could also postulate an L0 model, which would correspond to either endpoint of the L1 model, but this "monad" turns out to be trivial and not worth considering—at least from a mathematical point of view.) As previously mentioned, available computing power is finite, making it impractical to carry numerical investigations beyond the first few thousand stages of any model's development. This is particularly true with the L2 and L3 models, whose sizes increase in proportion to the square and the cube of time respectively. But I think my investigations are extensive enough to show how these models continue to behave as they mature. Of course, our own Universe is much, *much* older than any time that can be modeled numerically, but there is no reason to believe that the principles underlying its operation today are any different from those applicable at the outset.

The principles underlying the L1, L2 and L3 models not only carry forward in time, they can also undoubtedly be extrapolated to higher levels. However, I will not consider these levels here—not just because their dimensions are hard to visualize or because the mathematics becomes too gnarly, but for a more troubling reason. Each level contains copies of the lower-level models, so for example L4 (the four-dimensional model) would contain as its boundaries five copies of L3. That means that if we live in a universe similar to L3, there will probably be (on some hyper-plane of existence) an L4-like universe which contains five copies of ours. Our Universe has three-dimensional inhabitants, but that universe would presumably contain four-dimensional beings.

We've all laughed at two-dimensional beings that act as if they're alive and sentient, but whose behavior is in fact predetermined. If four-dimensional beings exist in some L4-like universe, we would appear to them very much as cartoon characters appear to us. The thought that we may be mere toons on the edges of some higher-level universe is not one that I choose to examine here.

6. Level 1: Fareyland

Ever Decreasing Circles: The Complete Series
—BBC DVD collection (2007)

Geometrical construction: Ford circles

The Level One model can be constructed in a number of ways. We will first look at a geometrical construction that uses a system known as the *Ford circles* (named after Lester Ford (1886–1967), the American mathematician who discovered them).

Take a piece of paper and draw a straight line and two equal-sized circles that touch each other and the line, like this:

Notice the small, roughly triangular space in between the circles and the line. Fill this in with the largest circle that will fit in this space, like this:

Now you have two of these triangular spaces between the circles and the line. Again, fill each one with the largest circle that will fit, like this:

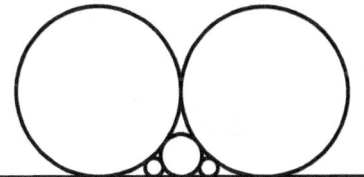

Now keep repeating this process over and over, at each stage taking the largest gaps along the line and fitting circles into them.

After you've done this a few times, you should have something that looks like this (enlarging the center of the diagram to show more detail):

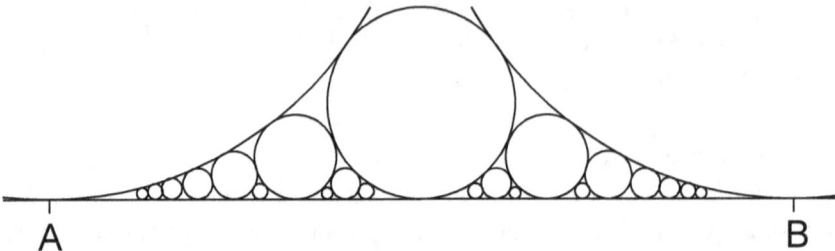

A   B

This symmetrical structure is called a system of *Ford circles* and as you'll see, it has some surprising properties and relationships.

The first surprise is that each circle's diameter divides exactly into the diameter of the starting circles. For example, if the starting circles each have diameters of one inch, all the other circles' diameters will be a unit fraction of an inch, like a quarter or a ninth. No complicated formulas involving pi, trigonometrical expressions or square roots. In fact, all the diameters are the reciprocals of perfect squares, like 1/4, 1/9, 1/16, 1/25 and so on.

The second surprise is that each circle touches the line at a point that is an exact fraction of the distance from A to B—again, none of these fractions will be Irrational numbers like square roots or pi. (The distance from A to B is the same as the diameter of the starting circles, and is the basic unit of distance for the model.)

But perhaps the most surprising feature is that if you keep filling in the gaps, eventually *any* point that is an exact fraction of the way along this line (called a rational point) will have a circle touching it. For example, after twelve stages there is a circle for every fraction of the distance from A to B, down to 1/12th, as this picture shows:

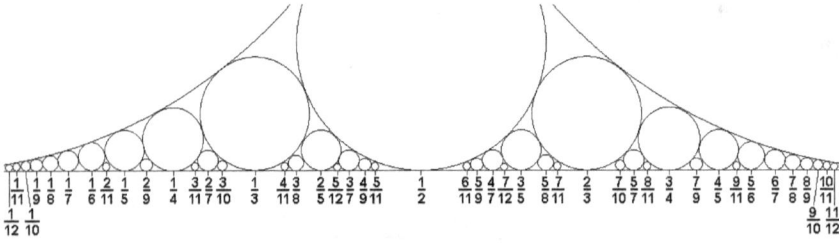

In other words, this construction eventually produces exactly one circle touching at each rational point along the number line. So equivalently, you can make this system by drawing a circle at each rational

point. Here's how: for any given rational point, we can say that it is p/qths of the way from A to B, where p/q is a fraction in lowest terms (meaning that p and q are Natural numbers with no common divisor greater than 1). Now draw a circle with diameter $1/q^2$ which touches the line at that point. This defines the same system of circles: one circle touches the number line at each rational point. Each circle (apart from the two starting circles at A and B) touches exactly two larger circles which we can consider its parents. Each circle also touches many smaller ones, but remarkably, *no two circles ever overlap*. (Of course, you could never draw *all* the circles, as there would be infinitely many, but you can draw them all down to a certain size, and you can choose this size to be as small as you wish.)

These circles are of various sizes, but rather than using the radius or diameter of a circle to indicate its size, it's more convenient to use the reciprocal of the diameter (1/diameter), which I call the circle's *order.* Thus, the circle at p/q has an order of q^2. Higher-order circles are smaller than lower-order ones and are generated later. In geometrical terms, a circle's order is equal to half of its curvature. And as already noted, all the Ford circles have orders that are perfect squares: 1, 4, 9, 16 and so on. There are no circles with other orders.

It can be shown geometrically that where two circles of radius a and b touch each other, the radius c of the child-circle generated between and touching them will satisfy the equation $\sqrt{a}\sqrt{b} = \sqrt{a}\sqrt{c} + \sqrt{b}\sqrt{c}$, as seen in this diagram:

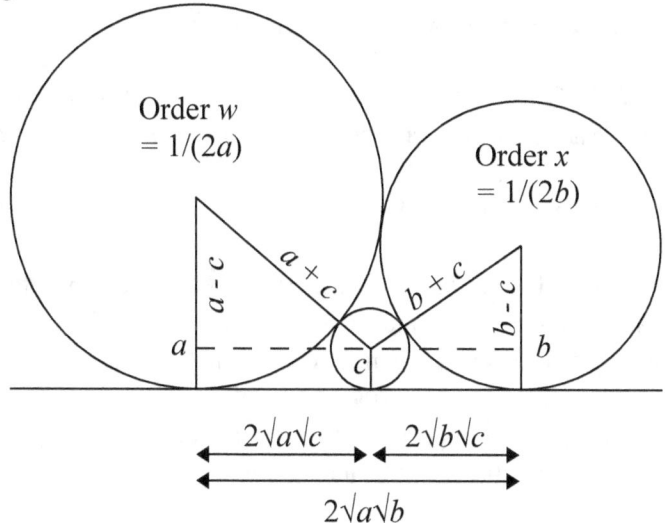

(The first two arrowed distances are derived by applying Pythagoras's theorem to the two right-angled triangles in the diagram, and the third one follows similarly.) Since each circle's order is the reciprocal of twice its radius, a little algebra shows that where two circles of orders w and x touch each other, the child-circle generated between and touching them will have order y, where $\sqrt{y} = \sqrt{w} + \sqrt{x}$.

This formula explains why all the circles' orders are perfect squares: their square roots are always obtained by adding whole numbers, starting with $1 + 1$ (the sum of the square roots of the orders of the two starting circles). Algebraically, the formula also implies that the orders w, x and y will satisfy the Cartesian equation $(w + x + y)^2 = 2(w^2 + x^2 + y^2)$, of which I will have more to say later.

In PQR Theory, this geometrical structure is known as L1, or the Level One model, because there is a single line that touches all the circles, and this line is a one-dimensional space. If you consider each point of this line as being created by the circle that touches it, then in effect the circles are creating the one-dimensional space. The circles determine the order in which the points of this space are created, but from now on, we will be more interested in the line than in the circles, since this line forms the "space" of our L1 model.

We will see in later chapters how this idea can be extended into two, three or more dimensions of space, but first let us look at other ways of building L1.

Numerical construction: Farey fractions

The whole fleet is lit up in fairy lamps, and each ship is outlined.... and the whole thing is in fairyland! It isn't true, it isn't here!
—Thomas Woodrooffe (1937)

The L1 model is not just a geometrical model. There are other ways to build the same structure. One such way is to use a series of fractions, known as the *Farey series* (named after John Farey (1766–1826), the English geologist who first described it).

The Farey series starts off with the numbers 0 and 1, which we represent as the fractions 0/1 and 1/1, like this:

0/1   1/1

Then we start to insert the intervening fractions, one step at a time—
first the halves, then the thirds, then the quarters, then the fifths
and so on, like this:

0/1			1/2			1/1
0/1		1/3	1/2	2/3		1/1
0/1	1/4	1/3	1/2	2/3	3/4	1/1

(note that we did not insert the fraction 2/4, as this is the same as 1/2)

0/1   1/5 1/4 1/3 2/5 1/2 3/5 2/3 3/4 4/5   1/1

0/1 1/6 1/5 1/4 1/3 2/5 1/2 3/5 2/3 3/4 4/5 5/6 1/1

… and continue this process indefinitely.

Note also that the fractions 2/6, 3/6 and 4/6 are not inserted, since
they are the same as 1/3, 1/2 and 2/3, which are already in the series. In
general, we only insert fractions in *lowest terms*: fractions which have no
common factor dividing both top and bottom. Interestingly enough, each
newly inserted term can be calculated by adding the tops and bottoms of
its neighbors. For example, 3/5 appears between 1/2 and 2/3, and is given
by the formula $\frac{3}{5} = \frac{1+2}{2+3}$.

It's easy to see that this numerical procedure produces the same
structure as the geometrical system of Ford circles described above, with
each fraction specifying a unique position along the line from A to B.

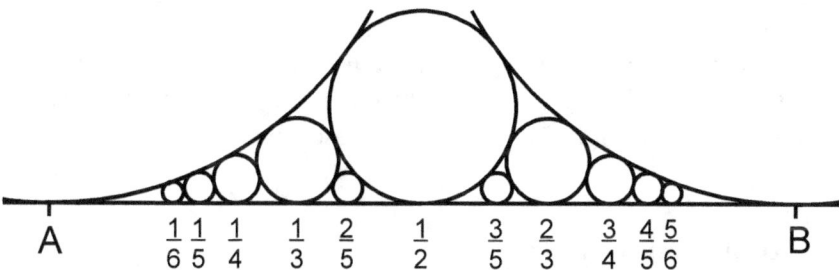

And if we continue this process indefinitely, the fractions become
closer and closer together, so that any given Rational number along the
line will eventually be included, just like with the Ford circles.

Another form of the Farey fractions: Duads

Let us now consider another numerical way of building the Level One model, and that is by using something I will call *duads*. (They should really be called dyads, but that term already has another meaning in mathematics.) These duads are not essential in Level One, but when we move on to Levels Two and Three, their higher-dimensional equivalents, known as *triads* and *tetrads*, will prove to be useful.

Let a and b be any two coprime Natural numbers (meaning that they have no common divisor greater than 1). Alternatively, a and b may be 0 and 1 in either order.

Let $m = a + b$ and let $y = m^2$.

Then the numbers a and b, in that order, are said to define the *duad* written as **a:b>y**. The number y is said to be the *order* of the duad.

There is a unique duad for every Farey fraction and vice versa. For example, the fraction p/q corresponds to the duad **p:(q−p)>q²** while the duad **a:b>y** corresponds to the fraction $\dfrac{a}{a + b}$. The reason is simple: if a and b are coprime, then so are a and $a + b$, and vice versa.

So for example, written in duad terms, the fourth row of the Farey series would be as follows:

0:1>1   1:3>16   1:2>9   1:1>4   2:1>9   3:1>16   1:0>1

corresponding to the following fractions:

0/1   1/4   1/3   1/2   2/3   3/4   1/1

This way of writing the Farey fractions illustrates the symmetry between the fractions to the left and right of the number ½: if you swap the order of a and b in any duad, you obtain its matching mirror-image counterpart.

Note that, like with the Ford circles, the duads of the L1 model have orders 1, 4, 9, 16, 25 and so on—all the orders are perfect squares. There are no duads whose orders lie in between these numbers.

An algebraic formulation

Four circles to the kissing come.
The smaller are the benter.
The bend is just the inverse of
The distance from the center.

Though their intrigue left Euclid dumb
There's now no need for rule of thumb.
Since zero bend's a dead straight line
And concave bends have minus sign,
The sum of the squares of all four bends
Is half the square of their sum.
—Frederick Soddy (1936)

In 1643, the French philosopher and mathematician René Descartes stated a geometrical theorem about the relationship between the radii of four touching circles. Expressed in terms of each circle's "curvature" (the reciprocal of its radius), this states that where four circles all touch each other, their four curvatures b_1, b_2, b_3 and b_4 satisfy the equation

$$(b_1 + b_2 + b_3 + b_4)^2 = 2(b_1^2 + b_2^2 + b_3^2 + b_4^2).$$

In 1936, this theorem was restated as above by the Nobel prize–winning radiochemist Frederick Soddy, in a poem called "The Kiss Precise." Soddy used *bend* to mean a circle's curvature, so the bend of a Ford circle is equal to twice its order. You can therefore substitute "order" for "bend" in the poem above, and it will still work (although it won't scan as nicely).

The poem also holds true in the special case where one of the four circles is a straight line (as it says, this is equivalent to a circle with zero bend). Thus for any *three* Ford circles that all touch each other and the straight line, the sum of the squares of their orders is half the square of their sum. In algebraic terms, their orders w, x and y satisfy the Cartesian equation $(w + x + y)^2 = 2(w^2 + x^2 + y^2)$ that I mentioned earlier. (*Cartesian* simply means "relating to Descartes.")

Like we saw with Pythagoras's theorem, this equation also works in reverse—each solution of this equation corresponds to a Ford circle in the L1 system, as follows: Let $\{w, x, y\}$ represent a solution to this equation in lowest terms, namely any three coprime Natural numbers which satisfy the equation $(w + x + y)^2 = 2(w^2 + x^2 + y^2)$. (Recall that *coprime* means the three numbers have no divisor in common apart from 1.) We will call such a set of three numbers a *Cartesian trio*. We always use y to represent the largest of the three numbers, and we call this the *order* of the trio. However, w may be more or less than x: these two numbers are allowed to appear either way round.

So, for example, $\{4, 25, 49\}$ and $\{25, 4, 49\}$ would represent two different Cartesian trios of order 49. (But the numbers {8, 50, 98} do *not* form a Cartesian trio, even though they satisfy the equation, because they are not in lowest terms. They are not coprime because all three numbers are divisible by 2; when you remove the common factor, they reduce to the Cartesian trio {4, 25, 49}.)

If {w, x, y} form a Cartesian trio, it is simple to show that w, x and y have to be perfect squares which satisfy the relationship $\sqrt{w} + \sqrt{x} = \sqrt{y}$. And this is exactly the equation that describes two Ford circles of orders w and x touching each other and generating a child-circle of order y, as in the following diagram:

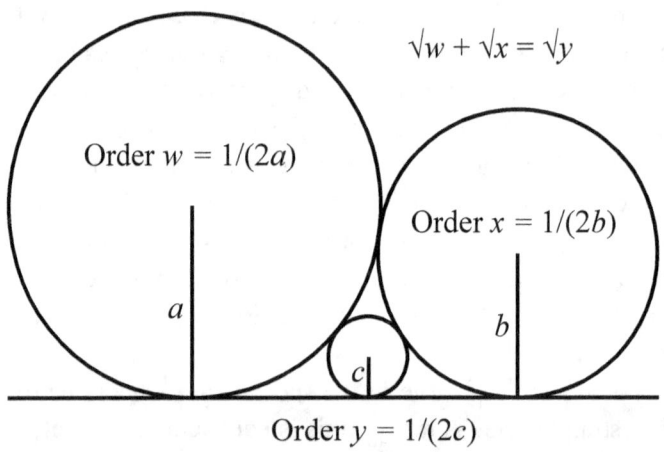

In this manner, each Cartesian trio corresponds to the generation of exactly one Ford circle in the L1 system. And if you swap the order of w and x, you obtain the Ford circle's mirror-image matching counterpart on the other half of the number line.

Four constructions, one structure

So far we have seen how the L1 structure can be obtained in four different ways: geometrically using Ford circles, numerically using Farey fractions, numerically using duads, or algebraically using Cartesian trios, the solutions of the equation $(w + x + y)^2 = 2(w^2 + x^2 + y^2)$. (This type of equation, which has to be solved using whole numbers, is known as a Diophantine equation.) And although they're four different constructions, the resulting structure is identical in each case. This chart illustrates the correspondence between the four methods, for orders up to 36 (corresponding to the first six rows of the Farey series):

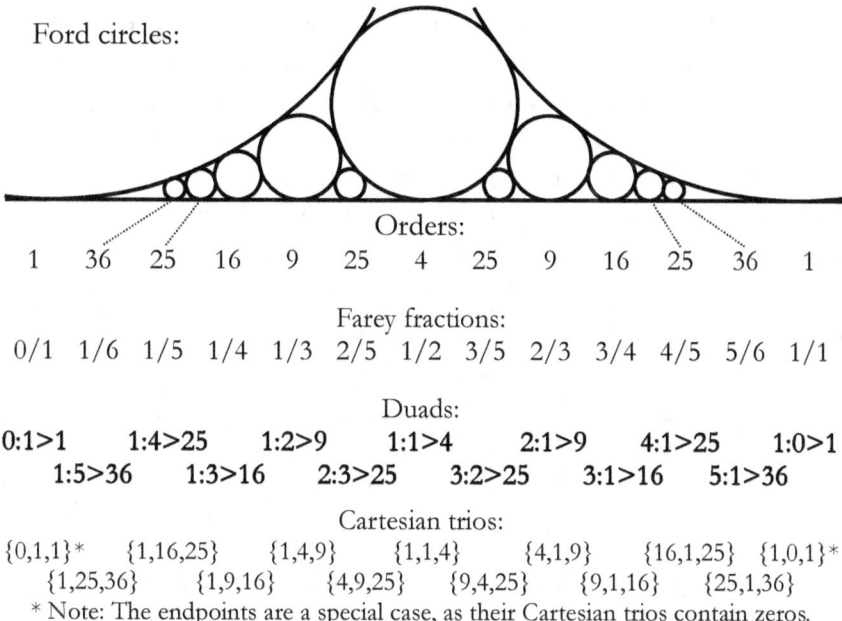

Ford circles:

Orders:

| 1 | 36 | 25 | 16 | 9 | 25 | 4 | 25 | 9 | 16 | 25 | 36 | 1 |

Farey fractions:

0/1 1/6 1/5 1/4 1/3 2/5 1/2 3/5 2/3 3/4 4/5 5/6 1/1

Duads:

0:1>1   1:4>25   1:2>9   1:1>4   2:1>9   4:1>25   1:0>1

1:5>36   1:3>16   2:3>25   3:2>25   3:1>16   5:1>36

Cartesian trios:

{0,1,1}*   {1,16,25}   {1,4,9}   {1,1,4}   {4,1,9}   {16,1,25} {1,0,1}*

{1,25,36}   {1,9,16}   {4,9,25}   {9,4,25}   {9,1,16}   {25,1,36}

* Note: The endpoints are a special case, as their Cartesian trios contain zeros.

The physical interpretation: Space and time

Each of these constructions creates the same series of rational points along the number line between 0 and 1. Each point has an order associated with it, which we will consider to represent the "time" at which it was created. So initially, Fareyland can be considered as starting with just its two endpoints, but as the time progresses and the model evolves, more and more points are created. In effect, we are creating a one-dimensional space, one order at each time, which becomes more and more finely grained at each stage of the construction.

Note that in this model all the orders are perfect squares, so that the only "times" when anything happens are perfect squares, namely 1, 4, 9, 16 and so on. In this particular model, nothing happens in between these times, but in later models we will find things happening more frequently.

We can carry on this process indefinitely, at each stage increasing the time and subdividing the space. Although we'll never reach infinity, we will eventually reach any finite number of steps you might care to mention. Each new stage can be considered as representing the passage of a quantum of time, and also as representing a subdivision of the space along the unit line. And after enough time has gone by, the points along the line will be so close together that we cannot see the gaps between them: it will seem like a continuous space, even though it isn't one.

The following chart shows how the number of points in this model increases at each stage. (Since the two endpoints can be considered as half-in and half-out of the space defined by the model, they only count as one point, not as two. Or you can consider the two endpoints as being one and the same, in which case the model "wraps around" on itself.)

Stage (q)	Order or time (t)	No. of points (n)	n/t	Stage (q)	Order or time (t)	No. of points (n)	n/t
1	1	1	1.0	12	144	46	0.31944
2	4	2	0.5	20	400	128	0.32
3	9	4	0.44444	50	2,500	774	0.3096
4	16	6	0.375	100	10,000	3,044	0.3044
5	25	10	0.4	200	40,000	12,232	0.3058
6	36	12	0.33333	500	250,000	76,116	0.30446
7	49	18	0.36735	1,000	1,000,000	304,192	0.30419
8	64	22	0.34375	2,000	4,000,000	1,216,588	0.30415
9	81	28	0.34568	5,000	25,000,000	7,600,458	0.30402
10	100	32	0.32	10,000	100,000,000	30,397,486	0.30397
11	121	42	0.34711				

As this chart shows, the number of points grows in approximate proportion to the time. The last column shows that as the time becomes large, this proportion settles down to approximately 0.30396, which is $3/\pi^2$. (Note for number theorists: this is because the number of points added at each stage is given by Euler's totient function.)

Energy and matter

Having thus defined the space and time in L1, let's now see how energy and matter can fit in. Remember, the space consists of a series of points strung out along a line, and as time passes new points are formed in the gaps between them.

Consider any two points along this line that are adjacent at some time (or, equivalently, consider any two Ford circles which touch each other). Sooner or later, a new point will appear between these two points, after which they will no longer be adjacent. Equivalently, if we look at the Ford circles, the two touching circles will spawn a third child-circle between and touching them both.

There is another way to view this process. Instead of considering the *points*, we can see what happens to the *line segments* that join adjacent points. And of course when a new point forms between them, the line linking them is destroyed and replaced by two line segments, like this:

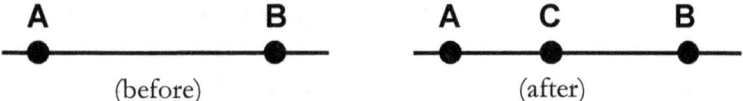

We can regard this process as a matter/energy interaction. Each point represents a matter/energy interaction occurring at a specific place and time; the lines joining these points represent the transfers of matter/energy between these interactions. (Remember, we consider matter/energy to behave like matter particles when it interacts but like energy waves while it is in transit between interactions.) Let us assume that A, B and C in the diagram above represent points in L1 having orders of a, b and c respectively, with $a < b < c$. In other words, A was formed before B, and B before C. The first picture represents the situation during the interval between times b and c, while the second picture shows the situation after time c. In the first picture, the line AB represents a quantum of "energy in transit" that inhabits the gap between A and B (it does not have a more definite position). At time c, an interaction occurs at position C, following which there are two quanta of energy in transit: one in the gap between A and C, the other in the gap between C and B.

The interaction at C has had the effect of splitting the quantum of energy represented by the line AB into two energy quanta, represented by the line segments AC and CB. (These new quanta will in turn subdivide again when new points form between A and C, and between C and B.) However, an observer detecting the two quanta could also treat them as evidence that a matter particle had decayed at position C at time c, and this would be an equally valid way of viewing the situation. (This alternate view is considered below.)

In order for our model to display a conservation law, we will say that the energy in each quantum is proportional to the length of the line segment it inhabits. Then the total energy in the model at any time will always be equal to 1. In other words, *energy will be conserved* during all these interactions.

The following diagram shows how the energy in the L1 system becomes subdivided during the first six stages of its evolution. The hollow dots represent the interactions (the points of the space which are newly formed at each stage), and the dotted lines indicate how the energy in each line segment subdivides at each stage. Arrows on the line segments indicate that the energy quanta are always considered as *moving away* from the point of the interaction at which they are created.

SUBDIVISIONS OF LINE ENERGY (UNSCALED)

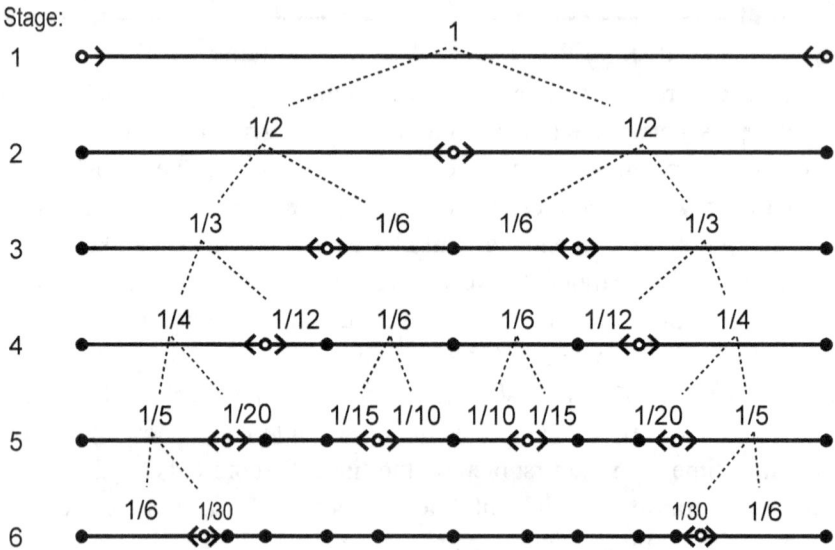

As you can see from the diagram, the initial energy is 1 at stage 1. At stage 2, it divides into two energy quanta, each with an energy value of 1/2; these in turn each subdivide at stage 3 into 1/3 + 1/6. At stage 4, the quanta of value 1/3 subdivide into 1/4 + 1/12, while the quanta of value 1/6 do not subdivide until stage 5, when they split into 1/10 + 1/15; and so on. The unlabeled line segments simply retain the values they had in the previous stage, so the total energy always remains at 1 in this physical interpretation. (Later on, we will consider what happens when we apply a scaling factor.)

Another way to depict the subdivision process graphically is by means of a Feynman diagram representing the decay of one quantum particle into two:

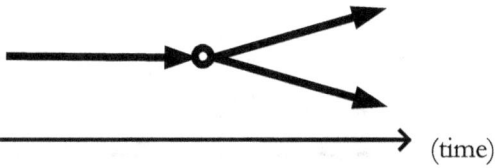

(time)

In the L1 model, this is the only type of interaction we see: a single quantum of matter/energy that subdivides into two similar quanta. So there is really no distinction between matter and energy in this model. You may be wondering whether it is possible to have different kinds of quanta (which would be represented by different kinds of line in a Feynman diagram) or whether one quantum can ever interact with another. This does not occur in L1, but we will see it happening in the higher-level models.

An alternate view: Point energy
Instead of regarding the energy as inhabiting the line segments joining the points, we could alternatively view it as inhabiting the points of space. In that case, we would regard each point along the line as containing an energy equal to *half* the distance between its nearest neighbors at any given stage of the process. This is illustrated in the diagram on the next page, where the points are labeled to show how much energy they contain after each of the first six stages. This approach gives the same energy-conservation principle we saw before: the total energy in the system is always equal to 1.

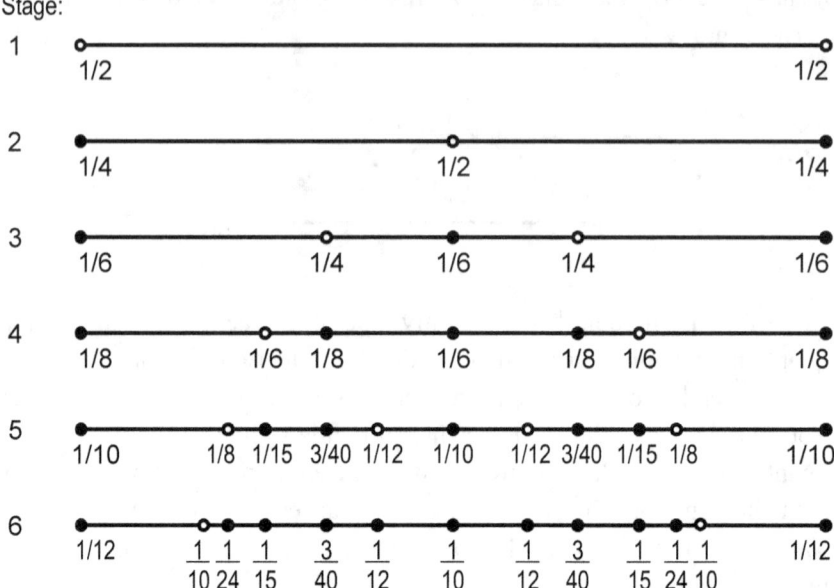

POINT ENERGY (UNSCALED)

As before, the white dots represent the points that are newly created at each stage, but now these points are considered to *acquire energy from their neighbors* when they come into existence. Thereafter, they gradually lose this energy as they themselves acquire new neighbors. As before, each point of the model is associated with a single matter/energy interaction. This means that at any given point of the model's space, only one interaction ever occurs; once a point has been formed, no subsequent interactions occur there. The black dots in the diagram therefore represent granules of space where nothing will ever happen again, even though energy is present in those granules. Thus, this approach gives a "vacuum energy" resident in each point of "empty" space, similar to the background energy which cosmologists have hypothesized exists in the empty reaches of space.

In the next diagram, arrows have been added to show how much energy is transferred at each stage. So, for example, the two endpoints each start with energy of ½ at stage 1; at stage 2, they each contribute ¼ to the newly formed central point, which now has total energy of ½ while the endpoints' energy is reduced to ¼ each. At stage 3, new points are formed one-third and two-thirds of the way along the line and acquire energy of 1/12 and 1/6 from their neighbors, and so on. Each point's energy is diminished by the arrows pointing away from it but increased by the arrows pointing toward it, so that total energy is conserved:

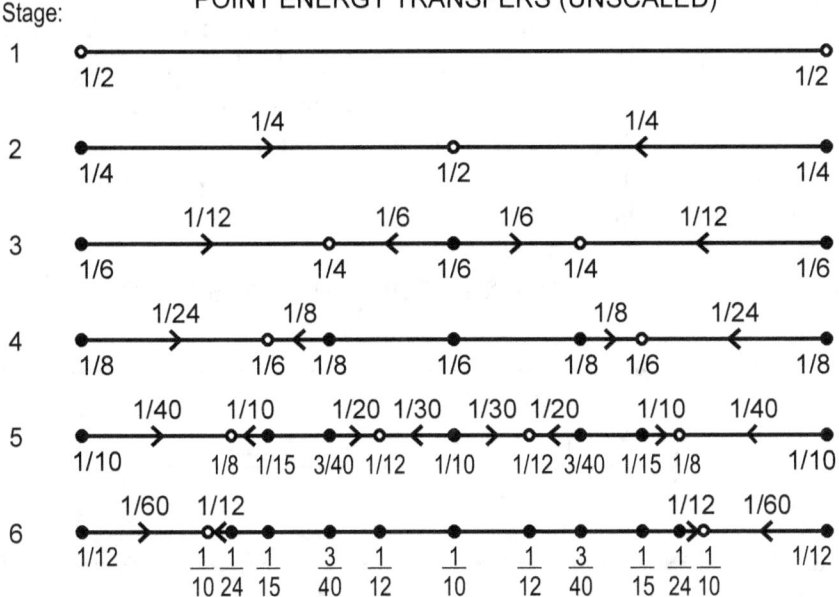

I wish to emphasize that these two views, line energy and point energy, are alternative interpretations of the *same underlying process*; neither is more valid than the other. Indeed, they can be considered as "dual" to each other: there is a line segment joining any two adjacent points, and a point between any two adjacent line segments. This means that one can switch back and forth symmetrically between the two inter-pretations, because the elements of one are the links of the other. This dual approach is analogous to the wave–particle duality that physicists have learned to accept when considering the behavior of matter/energy at the quantum level. A quantum behaves *both* as a matter particle *and* as an energy wave; it cannot be treated exclusively as one thing or the other, since its apparent nature depends on what experiment is being per-formed. It is the same with the L1 model: the apparent nature of matter/energy depends on how you choose to look at it.

Scaled energy

Now let us look at what happens under these two alternative views when we apply a scaling factor to the model—instead of space remaining the same size but becoming more finely grained as time goes by, we will now consider space as expanding but the granules remaining the same size. Under this alternative approach, the total energy in the system will grow over time.

To start with, we will simply use the number of the stage as our scaling factor, so that the original energy is doubled at stage 2, is tripled at stage 3 and so on. Thus, at any point in time, the total scaled energy in the system will be equal to the stage number. (This approach still ensures that energy is conserved *at each stage*, while it grows in between the stages.) First, the line energy:

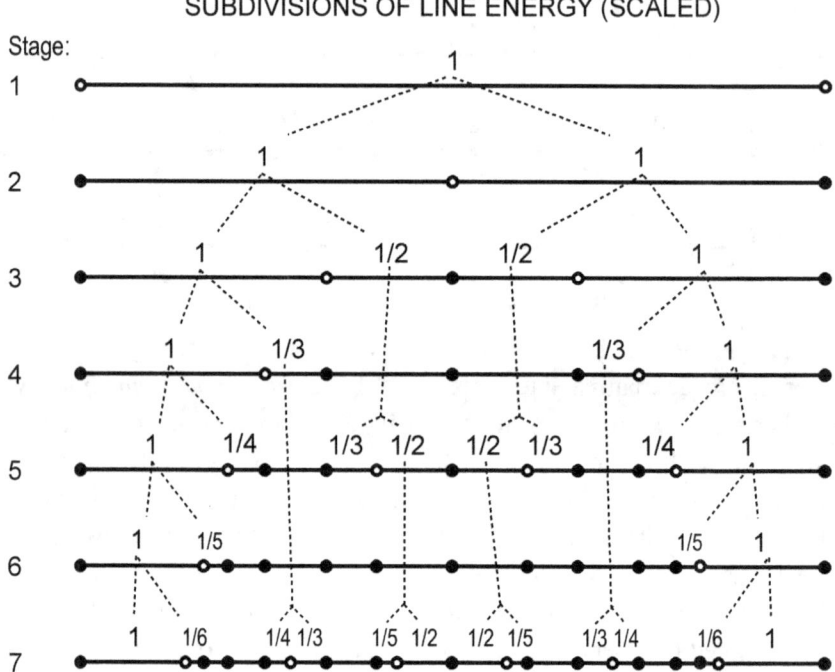

This diagram has been extended to include an extra stage so that you can see a pattern here: energy quanta which spread out to right and left, remaining unchanged as they go, and which periodically spawn new quanta that branch off to move in the opposite direction. A similar pattern appears when we consider the transfers of point energy:

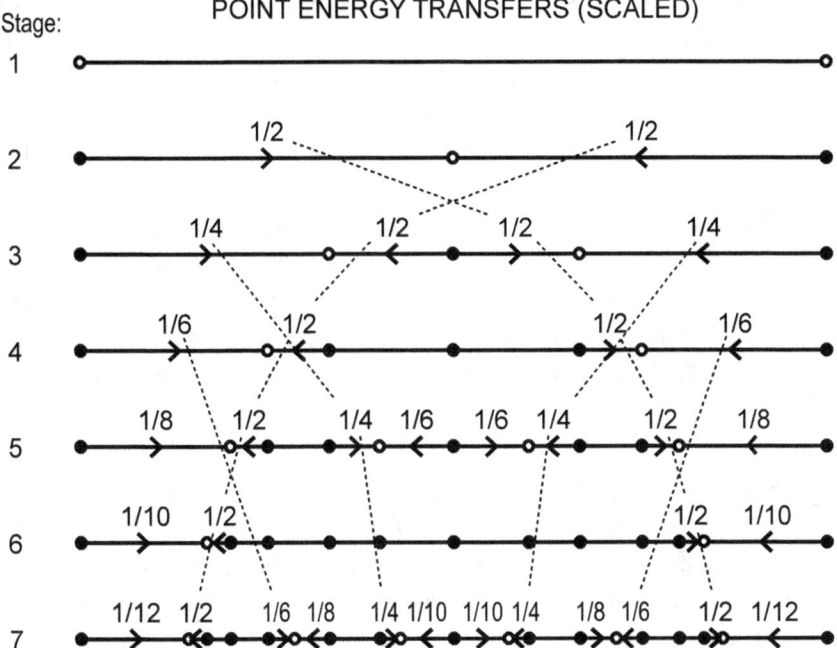

Here, the dotted lines show energy moving through the system, both to right and to left, with its scaled value remaining constant. For example, at stage 2, scaled energy of ½ moves from the left endpoint to the center point; at stage 3 it moves on to the two-thirds point; at stage 4 to the three-quarters point, and so on—a connected series of energy transfers. As we will soon see, these transfers seem to behave somewhat like photons.

The view from inside

So far, we've been looking at the L1 model from our external perspective of space and time. Indeed, in order to build the geometrical model, we utilized our preexisting ideas of space and time. However, if we really want to build a universe from scratch, we should not rely on these concepts; that would be circular reasoning (in more ways than one). Fortunately, the numerical and algebraic constructions do not suffer from this problem.

Instead, they raise another question: if we are not allowed to assume any preexisting ideas of space and time, how can we realize these concepts *within* the model? In other words, if this model universe had inhabitants, how would they perceive their space and time? It turns out that their perspective is going to be somewhat different from ours. From our viewpoint outside looking in, we can see the big picture and get a cosmic view of space and time in this model universe. But its inhabitants don't have that luxury: all they can experience is the space and time *where they are*. And those concepts may appear to operate differently for different inhabitants.

We experience this problem in our own Universe when we consider the effects of special and general relativity, of which I will say more later. For now, it is enough to say that time and space are relative to the observer; different observers will see time and space differently. However, the effects are only significant at astronomical scales—if two observers on Earth compare notes, they will find that any differences are minuscule.

Knowing this, how do we define time in our Universe? We don't have access to a cosmic clock that will give us some absolute, Universal time; all we can measure is the passage of time in our immediate neighborhoods. And while we can measure it very accurately, there is no general consensus on what time actually *is*. So I will simply be pragmatic and say that *time is what clocks measure*. And since clocks in different places may measure time differently, this means that time is a local phenomenon.

How do clocks measure time? Well, whether they use a pendulum, a vibrating spring, a quartz crystal, an atomic fountain or simply the trickling of sand, they all rely on physical processes of one kind or another. The rate at which these processes occur controls the clock's measurement of time. And these physical processes, in turn, are driven by the interactions between matter and energy.

In our model, as we have seen, the matter/energy interactions occur whenever the space subdivides to form new granules. So putting all this together, we find that the measurement of time in any locality will be determined by the formation of new granules of space. In other words, *the passage of time is the subdivision of space.* This is a key concept in PQR Theory and in understanding how our models can incorporate physical processes.

So although the inhabitants of the L1 model might not be able to detect the subdivision of their space directly, they could in theory detect it indirectly if they could build clocks. (Of course, this is all strictly hypothetical, as it is hard to see how a one-dimensional universe could contain clockmakers. However, the principles underlying the time and space of the L1 model are the same as for the higher levels. So although these concepts are discussed here for simplicity's sake, they are equally applicable to the higher dimensions.)

Now, how would these creatures measure space? Again, they would not be able to do it directly, as they cannot observe the points that make up their space. Instead, they would need to use instruments of some kind, such as measuring rods. From our external perspective, the length of such an instrument would depend on the spacing between its constituent particles, so (if we don't use a scaling factor) it would appear to shrink as the granules of the space become smaller.

But the creatures of that universe would not want to admit that their space was continually subdividing and becoming ever more finely grained. That approach would imply that their rulers would always be getting shorter (as would their subjects) and also that their physical "constants" would constantly be changing. Instead, like us, these people would undoubtedly find it more convenient to say that they, and all the things around them, are staying the same size while their universe expands. So in the last analysis they would be measuring their space in terms of the number of granules it contains at the time of measurement. And since their space is one-dimensional, they would only be measuring distance (not area or volume), which they would in effect be doing by counting the number of points comprising that distance. So to see things from their point of view, we need to use the model's order as a scaling factor (since the number of points increases in proportion to the order).

So now we have an idea of how time and space could be measured *locally* within a small region of the L1 model. The amount of space is just the number of points contained within that region at any given time, and the amount of time that appears to pass in an interval just depends on how often that number of points increases.

POINT ENERGY TRANSFERS, SHOWING SCALED WAVELENGTHS (SINGLE DIRECTION ONLY)

The diagram opposite shows how the transfers of scaled point energy appear to behave during the first twenty stages of the L1 model, if we measure space as the number of points. Accordingly, the diagram has been redrawn so that all the points at each stage are equally spaced. For clarity, only transfers moving to the right are shown, but there is a mirror image of transfers that move to the left: if the entire pattern were drawn, it would be symmetrical, as in the previous diagrams. Also for clarity, the numbers on the diagram are *half the reciprocals* of the scaled energy, so for example, the number 2 in this diagram corresponds to a scaled energy transfer of ¼ in the previous one. As before, dotted lines have been drawn to link these numbers, showing how the energy transfers move through the system, and how they appear to spread out at a more-or-less uniform speed. (Note that because the space is expanding, some of these transfers appear to be moving to the left; this is because we have chosen to fix the central point of the model at the center of the diagram. As a result, the points in the left half of the model are displaced to the left as one moves down through the diagram. However, as the previous diagram showed, each energy transfer actually departs from the same point at which its predecessor arrived, and in the same direction.) Again for clarity, only the dotted lines linking the lowest numbers are drawn, but similar lines can be drawn linking the larger numbers. These dotted lines represent energy particles which behave a lot like photons, as we will now see.

Some interesting observations can be drawn from this diagram. First, at each stage, the energy transfers have all the different numbers that are coprime with and less than the stage number. Thus, at prime-numbered stages, they include every number up to the stage number. For example, at stage 11, the rightward transfers are 1, 2, 3, 4, 5, 6, 7, 8, 9 and 10, and similarly valued transfers (not shown) are proceeding to the left. (This is a natural outcome of the way we built our model.)

More significantly, as you go down each dotted line the different numbers appear with different frequencies: the 1s appear at every stage, the 2s appear every two stages, the 3s appear every three stages and so on. Since the 1s represent scaled energy of ½, the 2s represent scaled energy of ¼ and so on, this means that each particle's **scaled energy is equal to half the frequency** with which it appears!

Why is this so significant? Because it reflects the Planck relationship $E=hv$ of quantum mechanics, where E is energy, v is frequency and *h* is Planck's constant, which happens to be 0.5 in this physical interpretation of the L1 model.

Now let us consider the speed at which, for an observer in this universe, these energy particles would *appear* to move. If this hypothetical observer were at a particular point, he could only detect the passage of time when the space in his immediate vicinity subdivided. These events would constitute his "clock ticks" or units of time. And each such event would involve an energy particle moving from his location to an adjoining point of space—in other words, one unit of space per unit of time. This means that our observer would measure the speed of light to be a constant 1 in the units we have chosen. And since the speed of light equals wavelength times frequency, this means that the numbers in this diagram can be considered as the wavelengths of the energy transfers they represent.

"This is all very well," you may say, "but what about the bigger picture? Do all the energy particles appear to travel at the same speed when we look at larger chunks of time and larger regions of this space?" I would answer that in this interpretation of the model, the energy particles moving to right and left keep on going forever, but they never catch up to each other. So we can say that they're all moving at the same speed *as one another*. However, whether this speed appears to be constant is going to depend on how we have defined the passage of time.

Now, you may have noticed that the stages in our diagram are not spaced uniformly but are positioned in proportion to the *order* of each stage. And although the dotted lines diverge due to the expansion of space, the spacing of the stages makes the dotted lines appear straight, so that each energy particle appears to be moving at a constant speed. This of course reflects our "cosmic view" from outside the model, where we defined the time at each stage to be the order, that is, the square of the stage number.

From inside the model, the quantum of time or clock tick is simply going to be the interval between one stage and the next, because nothing occurs in the interim and so no time can appear to pass. So the natural measurement of time will be the stage number. And because of the way this model is constructed, the "clock" ticks more or less uniformly across all its regions. But does this give the hypothetical inhabitants of that space a light-speed that stays constant? At most of the early stages (such as those shown in the diagram), the space expands by a large percentage upon each clock tick, and the effects of this expansion swamp the motion of the energy particles. But as the model matures, the expansion becomes less significant percentagewise, so that we can meaningfully trace the "distance" traveled by each energy particle during one cycle of its progress. For example, a particle with energy of 1/10, which is labeled as 5 in the diagram, appears once (that is, it goes through one cycle) every five stages of the model. The inhabitants of this land would say that its "frequency" was 1/5 and its "period" was 5. Using this terminology, they would find that the "distance" in points traveled by an energy particle during one cycle is given by the formula:

$$distance = \lceil (period + 1) / (originating\ stage - period) \rceil$$

(the square half-brackets denote rounding up to the next higher integer).

For example, the particle with a period of 5 that appears at stage 7 in our diagram appears again at stage 12. At this time it is three points to the right of where it previously appeared; and this distance is equal to $(5 + 1)/(7 - 5)$. It reappears again at stage 17, but now it is only one point to the right of the point at which it appeared at stage 12, which is because $(5 + 1)/(12 - 5)$, rounded up to the next higher integer, is 1. (Looking at the diagram you might think that it's moved more than one point to the right, but that's because the diagram has been drawn to reflect the expansion of space: each point moves away from the center as the stage number increases.)

Since speed is distance divided by time, the Fareylanders would find that for a particle of period T originating its cycle at stage q, its speed was given by the following formula (ignoring rounding):

$$speed = \left(1 + \frac{1}{T}\right) \times \left(\frac{1}{q - T}\right)$$

(note that T is always less than q, due to how the model is constructed). These inhabitants would be faced with a conundrum when trying to measure the speed of light, because each particle travels slower and slower as time goes by. As a result, the particles don't progress very far in their expanding space. Our example particle moves from position 1/2 to 4/7 at stage 7, from 4/7 to 7/12 at stage 12, and from 7/12 to 10/17 at stage 17. It is always approaching, but never reaches, position 3/5. The astronomers of this world might say there was a "black hole" at this position. (Indeed, they might notice that *every position* behaves like a black hole in this strange universe.) The problem arises because (in terms of cosmic time) their space expands at a uniform rate, but their clocks tick slower and slower as their universe ages. (This problem disappears by the time we get to the Level 3 model.)

And the physicists of this world might also notice something strange going on: based on the scaling factor that we chose, the average energy per point of space would be decreasing. This is because we chose to scale up the energy by using the stage number, whereas the number of points in the space grows with the *square* of the stage number. So these physicists would probably want to redefine their energy units in such a way that their physics remained as consistent as possible. In particular, they would want their basic quantum of energy (the content of the smallest particles of matter/energy) to remain constant over time. These particles' *unscaled* energies are inversely proportional to the square of the stage number (that is, the order or time). As a result, the physicists would presumably adopt a system of units that implicitly or explicitly used the square of the stage number as their scaling factor.

This change of approach has several consequences. First, the energy of each particle would appear to increase as it moved through the model. However, the effect on an individual particle would be immeasurably small by the time the model had passed through a few gazillion stages. Furthermore, the median energy of all the particles at each stage would remain constant.

Second, the "Planck's constant" within this universe would no longer be a constant but would increase in proportion to the stage number. Again, this effect would be minuscule after many stages.

And third, the *apparent* speed of a particle at any stage would depend on its energy content: the particles with the lowest energies would travel the fastest. At each stage, the fastest, least energetic particles would have a speed close to 1, using the measurement system I have described. (In fact their speed would be slightly higher than 1, but once again the error becomes vanishingly small as the stage number increases.)

Now these physicists would be faced with a philosophical puzzle: *why* do the more energetic particles travel more slowly? Many of them might be content to call this a law of nature. But one of them might come up with the crazy notion that the particles really all travel at the same speed, but that *the presence of matter/energy slows the passage of time.*

This would be an equally valid viewpoint—and of course, this time-dilation phenomenon is a well-established effect of general relativity.

But a note of caution is in place here. Our physical interpretations of the L1 model are somewhat artificial, and the formulas which govern how time is slowed by matter/energy in L1 are somewhat different from those which apply in general relativity. But while they are not definitive, these interpretations do serve to illustrate how some of the physical principles involved will operate in higher dimensions.

So now we have seen how our Pythagorean L1 model can, in principle, support physical interpretations corresponding both to the quantum and the relativity theories. Of course this model, being only one-dimensional, is very limited. There is little motion involved, and all the causal interactions are at short range. (In relativistic terms, one would say that the future light-cone of every event is very narrowly bounded.) Phenomena such as momentum and gravitational attraction appear to be absent (in fact, if anything, gravity seems to be a very weakly *repulsive* force in this model). Nor do the interpretations appear to support electro-magnetism or any differentiation in the types of matter/energy involved. We will have to look for these phenomena in the higher-level models, to which we now turn.

7. Level 2: The Flatrack

Time in hours, days, years,
Driv'n by the spheres.
—Henry Vaughan (1622–1695)

Geometrical construction: Ford spheres

Like Fareyland, the Level Two model can be built up in several ways. These methods correspond to the constructions described in the previous chapter, extended to incorporate a second dimension of space. So for our geometrical construction, instead of starting with two circles resting on a line, we'll begin with three spheres resting on a flat surface.

These three spheres are equal in size and touch each other to form an equilateral triangle. Into the central space of this triangle we can fit another sphere that touches them and also touches the plane surface. (If we imagine the spheres to be resting on a glass-topped table, this would be the view from underneath.)

This fourth, central sphere has one-third the diameter of the starting spheres and is surrounded by three gaps, into each of which we can nicely fit a sphere having one-quarter the diameter of the starting spheres, like this:

After several stages of this construction, the result will look something like this (enlarging the center to show more detail):

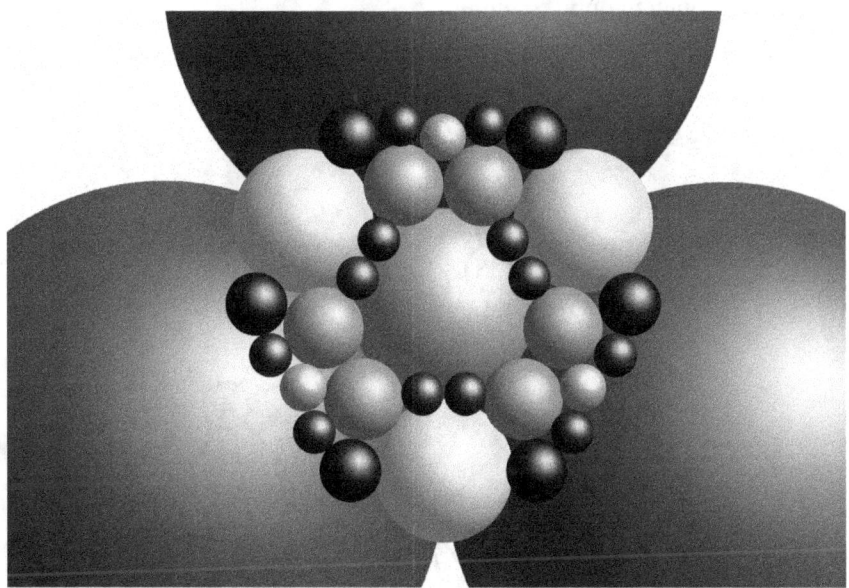

All the spheres are resting on the tabletop and touching their neighbors, and all have diameters which are exact fractions of the diameter of the three starting spheres. As we saw with Fareyland, if we assume the starting spheres to have diameters of 1, then each sphere's order is the reciprocal of its diameter. In the picture above, the spheres have orders 1, 3, 4, 7, 9, 12 and 13 (from largest to smallest), so they are considered as coming into existence at those "times."

This arrangement is known as a system of *Ford spheres* (which are nothing like Nissan Cubes). I call this model the Flatrack, because it vaguely resembles a triangular rack of balls on a pool table. (Of course, pool balls are all the same size and Ford spheres are not. And in practice, if you made this model using real balls of various sizes, you would build it upside down, placing the glass on top last so as to touch all the balls.)

Like in the previous level, if we continue this process of adding spheres long enough, one of them will eventually touch the table at *every* rational point in the original unit triangle. (These are the points which have rational coordinates in the skew system described below.) And these rational points, bounded by our original triangle, comprise the two-dimensional space of the L2 model.

In mapping out this model in two dimensions, it is much clearer and simpler not to draw the spheres themselves but just to mark the places where they touch the unit triangle:

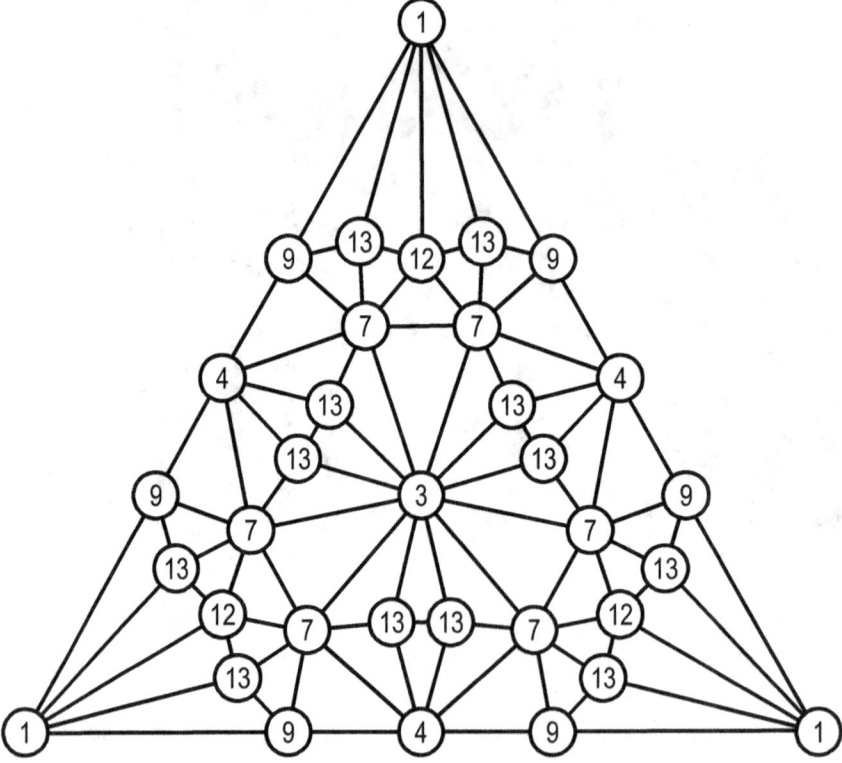

This diagram shows the L2 model as it exists after time 13 (the same time as in the previous picture). The circles show the points where the spheres touch the plane, and each circle is labeled with its sphere's order. The links joining these points indicate spheres that touch each other, and these lines divide the space of the diagram into triangles.

As you can see, the model has a six-way symmetry, as illustrated by the small diagram at right: the three lines of symmetry divide the model into six triangular sectors, each of which is a mirror image of its two neighbors. (As you may notice, there are a few places where two spheres of the same size touch each other across the lines of symmetry: these are called kissing sisters.) You may also notice that each edge of the model is a copy of the L1 model; at time 13, this is a line with five points of orders 1, 9, 4, 9 and 1.

In the L1 model, the orders of the elements were the perfect squares. In the L2 model, the orders of the elements are the Löschian numbers—numbers of the form $x^2 + y^2 + xy$, where x and y are integers. (About 23% of the first ten thousand Natural numbers are Löschian; the proportion gradually declines as the numbers get larger.) The first three Löschian numbers are 1, 3 and 4, so that the first three stages of the model occur at times 1, 3 and 4, as mapped out in these diagrams:

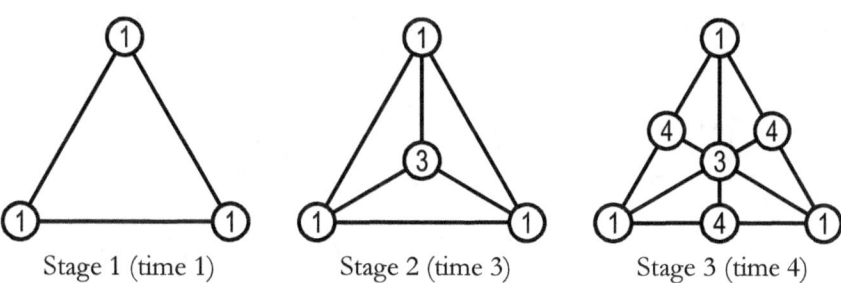

Stage 1 (time 1)   Stage 2 (time 3)   Stage 3 (time 4)

As you can see, at each new stage the triangles formed by the lines become subdivided as new points form. This process of subdivision operates in two different ways within the L2 model. The simplest way is when a new point forms inside a triangle, like this:

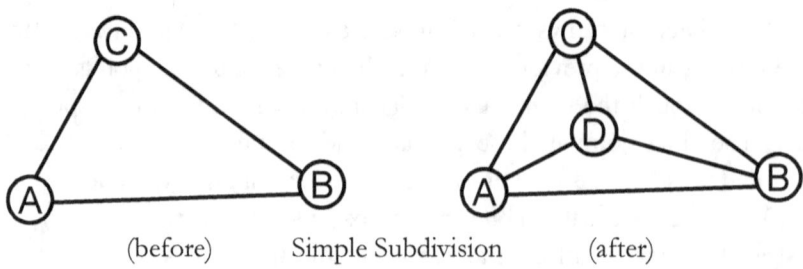

(before) Simple Subdivision (after)

Here, the formation of the point D within triangle ABC has subdivided it into three new triangles: ABD, ACD and BCD. In three-dimensional terms, this corresponds to a small child-sphere forming in the triangular space enclosed by three larger parent spheres, making four spheres all touching each other on a flat surface. Now, it can be shown geometrically that, for such a configuration, the spheres' orders v, w, x and y will satisfy the equation $(v + w + x + y)^2 = 3(v^2 + w^2 + x^2 + y^2)$. (I call this the Soddy equation—more on it soon.) Accordingly, the orders of points A, B, C and D in a simple subdivision will satisfy this equation. For example, a triangle formed by three touching spheres of orders 4, 7 and 9 will generate a fourth sphere of order 19, since $(4 + 7 + 9 + 19)^2 = 3(4^2 + 7^2 + 9^2 + 19^2)$.

The more complex way in which the model's space subdivides is called compound subdivision, and it occurs when a new point subdivides *two* adjacent triangles into *four*, like this:

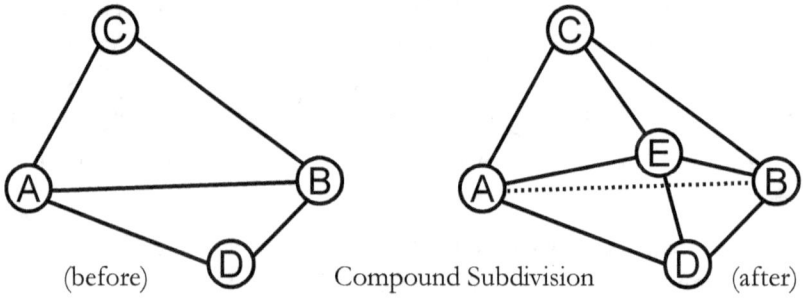

(before) Compound Subdivision (after)

Here, the formation of the new point E, close to the boundary between two adjacent triangles ABC and ABD, has subdivided the area occupied by these two triangles into four new ones: ACE, ADE, BCE and BDE. The border line AB between the two old triangles has been destroyed in the process. In compound subdivision, points A and B are always both lower in order than points C and D, so the link that is destroyed is always the oldest one involved. (In fact—as was the case in the L1 model—every link that is created is eventually destroyed.)

In terms of our three-dimensional model, compound subdivision occurs when two large spheres touch each other and are flanked by two smaller spheres, enclosing a four-sided gap. When an even smaller sphere forms in this space between and below the two large spheres, it "breaks the link" between them, as the following picture illustrates:

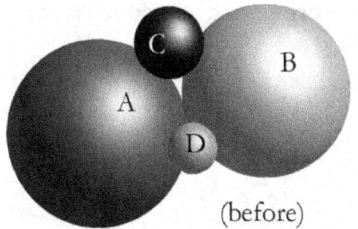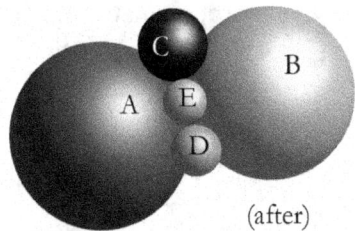

(before)   (after)

Thus, because their spheres are all touching each other, the points A, B, C and E in a compound subdivision will have orders that satisfy the Soddy equation, and so will the points A, B, D and E. For example, in our picture the spheres A, B, C and D have orders of 12, 13, 31 and 43 respectively. Thus, the order of the sphere E that forms in the middle will be 49, since

$$(12 + 13 + 31 + 49)^2 = 3(12^2 + 13^2 + 31^2 + 49^2) \text{ and}$$
$$(12 + 13 + 43 + 49)^2 = 3(12^2 + 13^2 + 43^2 + 49^2).$$

Geometrically, these two equations are saying that it's as if the two adjacent triangles ABC and ABD *each independently generate the same child-sphere* at E. No wonder the boundary between them dissolves!

The numerical construction

As with the L1 model, we can also build this structure numerically. But when we try to use the normal Cartesian coordinate system to build our triangular space, we are faced with a problem. If we use the coordinates (0,0) and (0,1) for the base of our triangle, then the apex will have coordinates ($\frac{1}{2}$,$\sqrt{3}/2$) and we will find ourselves using Irrational numbers. To avoid this problem, we need to use a non-orthogonal (i.e., skew) coordinate system where the axes are inclined at 60° instead of 90°.

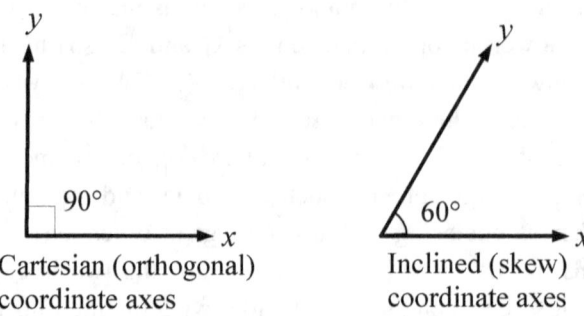

Cartesian (orthogonal)
coordinate axes

Inclined (skew)
coordinate axes

This skew coordinate system works much like the Cartesian one: any point's coordinates can be found by measuring its distance from each axis along a line parallel to the other axis, like this:

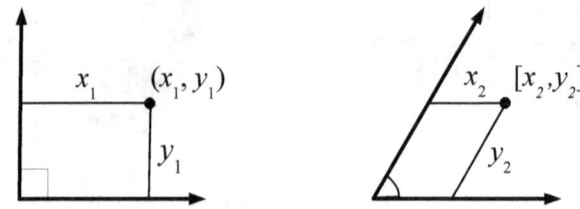

Here, the point with coordinates (x_1, y_1) in the Cartesian system has coordinates $[x_2, y_2]$ in the skew system, where

$$x_2 = x_1 - \frac{y_1}{\sqrt{3}} \quad \text{and} \quad y_2 = \frac{2y_1}{\sqrt{3}}.$$

Thus in this skew coordinate system our unit triangle would have corners at [0,0], [1,0] and [0,1], as shown in the diagram below. (The square brackets indicate that the skew system is being used.)

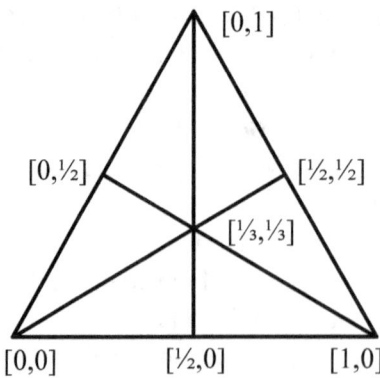

We can now say that the rational points in our unit triangle comprise all points $[x, y]$ whose (skew) coordinates x and y are non-negative Rational numbers with $x + y \leq 1$. And as these are Rational numbers, we can write $x = r/s$ and $y = t/u$, where r, s, t and u are four integers; in this way, each rational point can be specified by four integers. In fact, we don't need to use four integers here: we can represent a point's coordinates using only the three integers ru, st and su, since $x = ru/su$ and $y = st/su$. Let us reduce these three integers to lowest terms by eliminating any common factors, calling the results a, b and m respectively; then let $c = m - (a + b)$. In this way any point of the triangle having rational skew coordinates $[x,y]$ can be represented by the three integers a, b and c, where

$$x = a/(a + b + c) \quad \text{and} \quad y = b/(a + b + c).$$

The integers a, b and c will be coprime (as otherwise a, b and m would have had a common factor) and non-negative (as otherwise x or y would have been negative or $x + y$ would have been greater than 1).

I call such a set of three non-negative coprime integers a *triad*. (By the way, if two of the numbers in a triad are 0, then the third has to be 1.) Each triad defines a unique rational point in the unit triangle, and there is a unique triad for each such point. We can therefore use the numbers a, b and c as coordinates to represent these points. This representation is known as a system of *barycentric coordinates*. For the L2 model, this barycentric system, in which a point's coordinates are written as $a{:}b{:}c$, turns out to be preferable to using the skew coordinates and writing $[x,y]$.

For example, the three corners of our unit triangle have barycentric coordinates of 0:0:1, 0:1:0 and 1:0:0, and the center has coordinates 1:1:1. (The coordinates are called barycentric because they represent three weights which, if placed at the triangle's corners, would have their center of gravity or barycenter at the given point.) Note that in this coordinate system, the edge points have 0 as one of their coordinates, the corners have 0 as two of their coordinates, and points on the lines of symmetry have two (or at the center, three) equal coordinates. Note also that the order of a point's coordinates is significant: if you rearrange them, you will get the coordinates of a symmetrically matching point in another sector. The six possible permutations of the three numbers in a triad correspond to the six symmetrically matching sectors of our L2 space, as the following diagram shows:

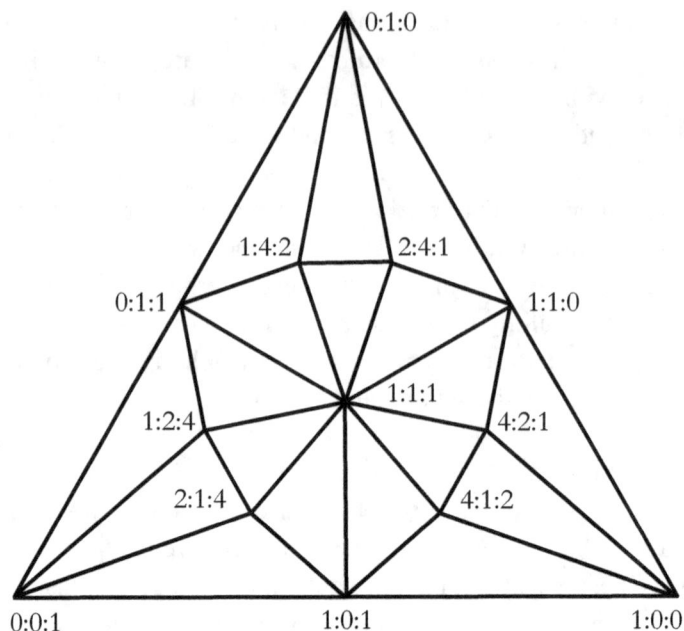

When we built the L1 model, we wrote our duads as **a:b>y**, and we can now see that *a:b* represented the barycentric coordinates of a point on the unit interval and that *y* represented its order. In the L1 model, *y* was equal to $(a + b)^2$.

Analogously, in the L2 model, we write the triads as **a:b:c>y**, where *a:b:c* represent a point's barycentric coordinates on the unit triangle and *y* again represents its order. But now the formula for *y* is more complex:

$$y = \frac{(a + b + c)^2}{\gcd(a^2 + b^2 + ab, \; a^2 + c^2 + ac, \; b^2 + c^2 + bc)}.$$

In this equation, *gcd* denotes the greatest common divisor of the three terms listed inside the bracket (i.e., the largest integer that divides evenly into all three terms). So, for example, the order of the point at 1:3:9 would be calculated as $13^2/\gcd(13, 91, 117)$. The largest integer that divides evenly into 13, 91 and 117 is 13, so in this case *y* evaluates to $13^2/13$, which is 13. Thus this point in the L2 model would be of order 13 and would be represented by the triad written as **1:3:9>13**.

Geometrically, the formula for *y* states that the diameter of the sphere at a given point is the largest amount that divides evenly into the *squares* of all three distances from that point to the corners of the space. That amount can thus be considered the model's basic unit of area at time y. Turning this around, this implies that the basic unit of area in the model varies inversely with the time. (In case you're wondering why the dimensions don't appear to match here, it's because the unit of area actually equals the sphere diameter times the length of the triangle side—but since that length is 1, it doesn't show up in the formula.)

Now, given any three coprime non-negative integers a, b and c, we can use this formula to calculate y, and y will always be equal to the order of the Ford sphere that touches the plane at the point $a{:}b{:}c$. And since every rational point in the unit triangle can be represented using non-negative integral barycentric coordinates, this means that each triad **a:b:c>y** corresponds exactly with one Ford sphere, and vice versa. In other words, this *purely numerical* construction of L2 has created the exact same structure as our geometrical method did.

An algebraic formulation

And now besides the pair of pairs
A fifth sphere in the kissing shares.
Yet, signs and zero as before,
For each to kiss the other four
The square of the sum of all five bends
Is thrice the sum of their squares.
—Frederick Soddy (1936)

In the last chapter, I quoted a verse from Frederick Soddy's poem "The Kiss Precise" giving the relationship between the "bends" or curvatures of four touching circles. In the next verse of the poem, Soddy extended his result from two dimensions to three, to give a formula relating the bends of *five touching spheres*. As you may recall, we are working with the special case in which one of the spheres is a flat plane of bend zero, and we're using orders that are one-half of Soddy's bends. So where four Ford spheres touch each other, the poem tells us that the sum of their orders, squared, is three times the sum of their squares, or, algebraically:

$$(v + w + x + y)^2 = 3(v^2 + w^2 + x^2 + y^2).$$

This is the Soddy equation I referred to earlier in this chapter. And just like the Cartesian equation discussed in the last chapter, this equation also works in reverse. Any four coprime Natural numbers fitting this equation are called a *Soddy quartet*, which we can write as $\{v, w, x, y\}$; as before, we always place the largest number last. It turns out that *every* Soddy quartet represents the orders of four spheres that touch each other in the L2 model. Thus the Soddy quartet $\{v, w, x, y\}$ represents a child-sphere of order y being generated by three parents of orders v, w and x. (The numbers in Soddy quartets are always Löschian.)

When we map out the L2 model in terms of triangles, we find an exact one-to-one correspondence between the Soddy quartets and the triangles that subdivide to generate the points of the model, as follows:

1. In most Soddy quartets, all four numbers v, w, x and y are different. After choosing y to be the largest number, there are six possible permutations of the other three numbers. There are also six symmetrically distributed triangles in the L2 model (one per sector) whose points have orders of v, w and x, and which subdivide at time y.

2. Some Soddy quartets contain a single repeated number, which may or may not be the largest of the four:

2a. If the repeated number is the largest, as in the quartet $\{1, 3, 7, 7\}$, this corresponds to the special case where two kissing sisters form across one of the lines of symmetry, as in the following diagram:

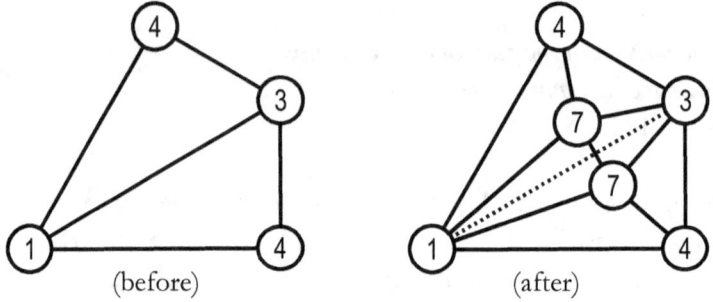

(before)　　　　　　　(after)

If we imagine that one of the two sisters is formed an instant before the other, this process breaks down into a simple subdivision immediately followed by a compound one:

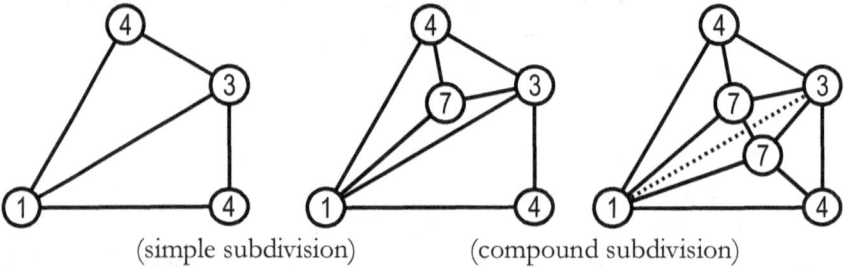

(simple subdivision)　　　(compound subdivision)

So here the Soddy quartet $\{1, 3, 7, 7\}$ corresponds to the long narrow triangle which is formed at the intermediate stage of this process but then immediately participates in the compound subdivision that forms the second of the two sisters. There are six arrangements of the numbers 1, 3 and 7, which correspond to the six kissing sisters (symmetrically distributed as three pairs) formed at time 7. (Note that there is a choice of which kissing sister we assume is formed first in each pair, just as there is a choice of which of the two 7s is the y in our Soddy quartet.)

2b. If the repeated number is not the largest, as in the quartets $\{3, 13, 13, 25\}$ or $\{7, 7, 12, 25\}$, this corresponds to a triangle formed by two kissing sisters and a third point along the line of symmetry between them. When the triangle subdivides, a new point is formed along this line of symmetry. In this case, there are three possible permutations of the numbers w, x and y, and three places in the model (one on each line of symmetry) where this subdivision occurs.

3. Finally, the Soddy quartet $\{1, 1, 1, 3\}$ contains three instances of the number 1, and this corresponds to the single center point of the model, where the three starting spheres generate a child-sphere at time 3.

The Soddy quartets don't just correspond one by one with the subdividing triangles. Any *pairs* of Soddy quartets that differ only in their next-to-largest members (pairs such as $\{13, 16, 37, 57\}$ and $\{13, 16, 49, 57\}$) correspond exactly with the *pairs* of adjacent triangles in the model that participate in a compound subdivision.

Thus, in a very real sense, the solutions of the Soddy equation form the same structure as that formed by the Ford spheres and by the triads. This is particularly significant because it shows how a purely *numerical* concept (the set of solutions to a particular equation) can also define a *geometrical* space—a space which evolves over time. And this, I believe, is also how our Universe is defined.

Toward a physical interpretation

In the L1 model, we saw how the number of points increased roughly in proportion to the order. In the L2 model, the number of points increases in approximate proportion to the *square* of the order, as shown in the following chart. (To avoid fractions, the edge and corner points are counted as 1 in this table, not as ½ or ⅓.)

Order or time (t)	No. of points (n)	n/t^2	Order or time (t)	No. of points (n)	n/t^2
1	3	3	20	61	0.15250
3	4	0.44444	50	343	0.13720
4	7	0.43750	100	1,249	0.12490
7	13	0.26531	200	4,900	0.12250
9	19	0.23457	500	29,803	0.11921
12	22	0.15278	1,000	119,020	0.11902
13	34	0.20118	2,000	473,650	0.11841
16	43	0.16797	5,000	294,759	0.11790
19	61	0.16898	10,000	11,778,742	0.11779

As you can see, over the long term, the ratio of the number of points to the square of the order settles down to be close to 0.1177. And since every triangle is bounded by three lines, and every line (except those along the edges) separates two triangles, the number of triangles will be roughly two-thirds the number of lines. (As the model matures, the elements along the edges become less and less significant as a proportion of the whole.)

Furthermore, we can consider the model as a planar graph (that is, a collection of points in a plane, connected by straight lines that don't cross). Then, by Euler's formula, the number of points plus the number of triangles equals the number of edges plus one.

Taken together, these relationships tell us that when the time t is large, the L2 model contains approximately $0.1177t^2$ points, $0.2354t^2$ triangles and $0.3531t^2$ lines. Each subdivision (whether simple or compound) creates one additional point, two additional triangles and three additional lines.

In the L1 model, we saw there was a duality between the points and the lines. We can extend this principle to the graph of the L2 system. If we draw a point in each region of the graph, and a line connecting points in adjacent regions, we get what is known as a dual graph.

The following diagram illustrates the graph of the L2 model at time 7, together with its dual (indicated by the dotted lines and hollow dots). The dual graph has a point in each region (including the exterior region) of the original graph *and vice versa*: the original graph also has a point in each region of the dual. Thus, the points of each graph correspond

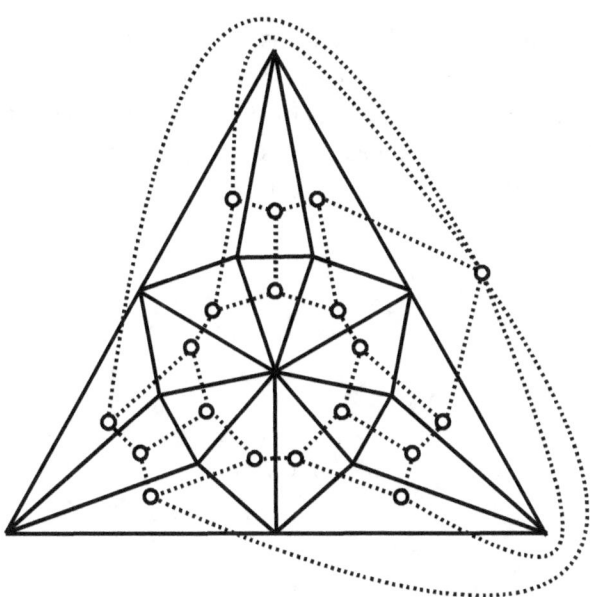

exactly to the regions of the other (so the dual of the dual is the original graph). Furthermore, each link in either graph crosses exactly one link of the other graph, so that the links of the two graphs also correspond exactly. (Note that in graph theory the positions of the points are not significant, only the connections between them. We could in principle rearrange the points of the dual graph so as to make all its lines straight.) In practice, when drawing the dual graph, it's tidier to omit the point corresponding to the exterior region and replace the lines leading to it with outward-pointing arrows, which are understood to meet up at infinity.

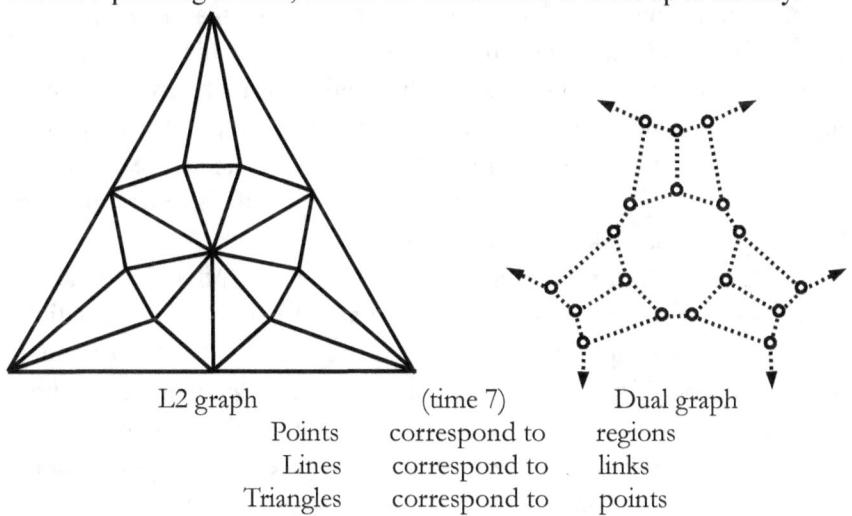

L2 graph   (time 7)   Dual graph

Points	correspond to	regions
Lines	correspond to	links
Triangles	correspond to	points

When we plot the dual graph this way, exactly three lines meet at every point, because the points correspond to triangles in the L2 graph. We can imagine these as the points of our space at a given time. And of course, as the L2 model evolves and its triangles subdivide, the points of the dual graph also subdivide correspondingly.

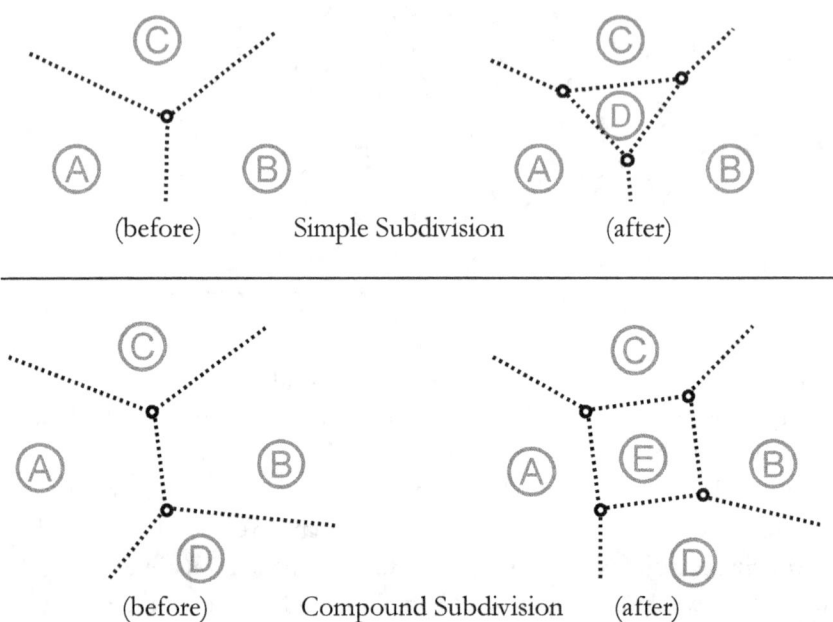

(before)　　　Simple Subdivision　　　(after)

(before)　　　Compound Subdivision　　　(after)

As these diagrams show, a simple subdivision replaces one point of the dual graph with three new points having three new links; a compound subdivision replaces two points and their link with four new points having four new links. Each subdivision (whether simple or compound) creates one additional region, two additional points and three additional links in the dual graph. (The regions are created with three or four sides but gain additional sides as time goes by.)

In general the links (apart from those pointing off to the exterior, which we will ignore from here onward) are of two types. Type 1 links are created in a simple subdivision or as the region C-E or D-E boundaries of a compound subdivision, and go on to participate in three subdivisions; type 2 links are created as the A-E or B-E boundaries in a compound subdivision, and go on to participate in two subdivisions. Both types are subsequently destroyed in a compound subdivision. So in effect, each link participates in a total of six subdivisions during its life-time. (Geometrically, this corresponds to the six spheres of a Soddy hexlet.)

Now, if we are seeking a physical interpretation, we can treat the points of the dual graph (apart from the point at infinity) as being the points of a two-dimensional space. This space has a natural measure of distance: at any given time, the distance between any two points is sim-ply the number of links along the shortest route joining them. Similarly, there is a natural measure of area, namely the number of points in a given section of the graph.

As with the L1 model, we can treat energy as residing either (1) in the points or (2) in the links of this space. Under the first approach, each point's energy would be the geometrical area of the corresponding trian-gle in the L2 model. Under the second approach, each link's energy would be one-third of the total area of the two triangles corresponding to its endpoints (but excluding the area outside the unit triangle). Under either approach, the total energy remains equal to the area of the unit tri-angle, so energy is conserved. And as before, either approach is an equally valid way of viewing the energy. (We could also treat the energy as residing in the dual graph's regions, which correspond to the points of the L2 model—an approach which will be useful when considering grav-ity.) Incidentally, it is convenient to work in *triangular* rather than square units of area, so that the total energy in our model (before we apply a scaling factor) will always be 1.

If we call the points of this space quarks and the links joining them gluons, then this model represents something akin to the quark-gluon plasma hypothesized by physicists. The entire space is filled with—indeed, consists of—these quarks and gluons. And because of the way we have defined distance, our gluons bind neighboring quarks together and keep them at a fixed separation.

If there were physicists living in this space, how would they per-ceive its matter/energy? Would they consider it as residing in the "quarks" or in the "gluons"? They would be faced with a conundrum, as the quarks can only be detected by their interactions with the gluons, and the gluons can only be detected by their interactions with the quarks. If the physicists believed their measuring equipment to be composed of quarks, then they would interpret it as detecting gluons—and vice versa. And both interpretations would be equally valid. (As Einstein told

Heisenberg: you cannot know what can be observed until you have a the-ory; only then can you define what can be observed.)

But either way, as we saw in the L1 model, these physicists would probably apply an implicit scaling factor to their energy quanta so that their physical laws would remain consistent. In this two-dimensional model, the natural scaling factor to use is the square of the order or time. Thus, at any given time t, each triangle's area would be multiplied by t^2 to give its scaled energy; the same scaling factor would also apply to the links. As before, this means that the total scaled energy in the system will increase over time, but this effect becomes less and less perceptible as the model ages, and the scaling allows the average energy content of these "quarks" and "gluons" to remain constant over time.

Distributions of energy

The following two bar charts show the distribution of scaled energy in this model at time 2000. First, the energy content of the triangles (rep-resented by their scaled area in triangular units):

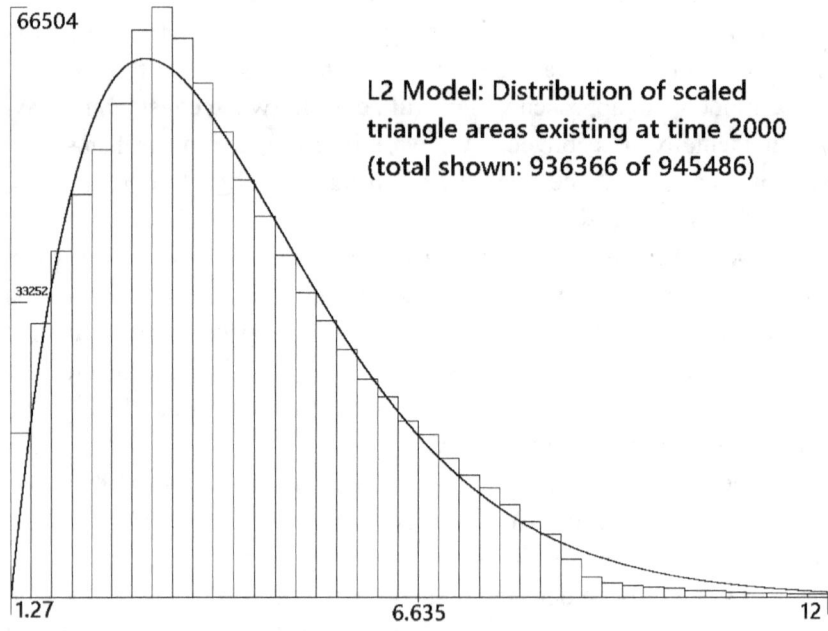

66504

L2 Model: Distribution of scaled triangle areas existing at time 2000 (total shown: 936366 of 945486)

33252

1.27   6.635   12

Second, the energy content of the links (represented by one-third of the scaled area contained in the two triangles they join):

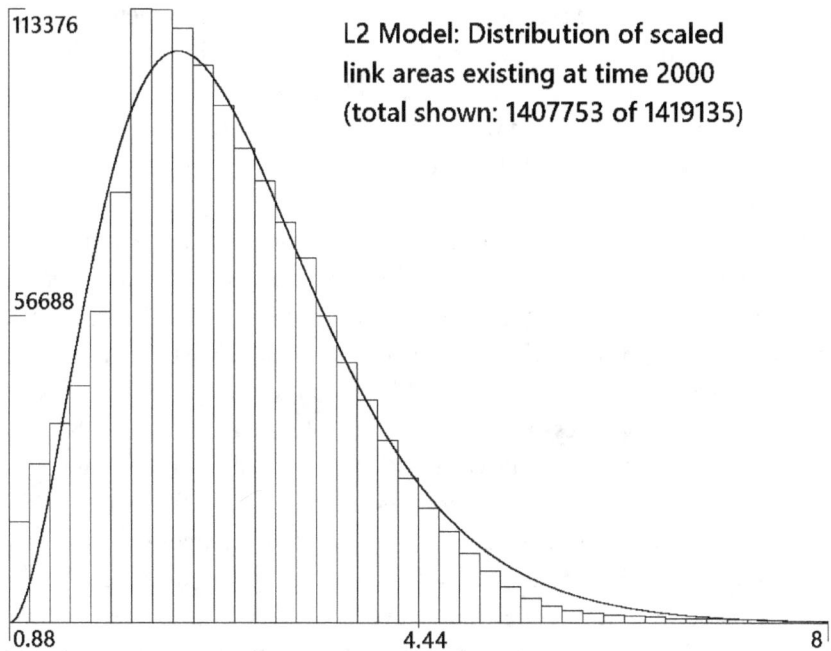

The end points of each chart have been chosen so that at most 0.5% of the triangles or links are cut off in each tail of the distribution; the remaining 99+% are included in each graph. As indicated on the vertical axes, the height of each bar reflects the number of triangles or links whose scaled areas (shown on the horizontal axis) fall within the range represented by the bar.

The curved line on each graph represents the "best-fit" Planck distribution curve, using the formulas $y = \dfrac{k\,x^2}{e^{\mu x} - 1}$ and $y = \dfrac{k\,x^3}{e^{\mu x} - 1}$ for the first and second graphs respectively, with the values of k and μ being chosen separately for each graph. (The curves were fitted to the bars using the method of least squares and the arbitrary constraint that they should start at the graph origin.)

The first formula, which models the *number of photons* as a function of the frequency x, was found to be the best fit for the first graph; the second formula, which expresses their *total energy* as a function of frequency, gave the best fit for the second graph. (The ratio between these formulas reflects the Planck relationship mentioned earlier: the energy of each individual photon is proportional to its frequency.) Although the fit for each graph is not exact at time 2000, it is significantly better than at earlier times, and there is every reason to believe that it will continue to get closer as the model time increases and the distortions attributable to its oldest elements become proportionately smaller.

The significance of all this is that as the L2 model matures, the distribution of its areas resembles the distribution of the energies in the photons emitted by an ideal thermal radiator (known as a black body) in theoretical physics. And the shape of the energy distribution curve for the links matches the black-body energy curve displayed by the cosmic microwave background (which cosmologists like to explain as the radiation left over from the Big Bang). This strongly suggests that at least some of the same processes are at work in nature as in our models. (But note that in the second graph we are comparing apples with oranges: the height of the bars reflects the *number* of photons at each frequency, while the cosmological curve represents their *total energy*. Hopefully, when we add a third dimension to our model this discrepancy will disappear.)

Incidentally, the cosmic microwave background is *almost* uniform in every direction, but with minuscule variations that are attributed to quantum fluctuations in the early Universe. The L2 and L3 models display similar quantum variations in their energy density as you move from place to place in the models. In a model that represents our Universe, it should be possible to find a place at which, as one looks in different directions, the energy pattern will match the cosmic background. This is an area where further study of the L3 model may prove fruitful.

A colorable argument

The graph of the L2 model displays another interesting property: we can color its lines, using only three colors, so that every triangle's three sides are three different colors. (A graph theorist would say that its medial graph is 3-colorable.) And once we assign colors to the three sides of the original triangle, this uniquely fixes the coloring for all the lines that form subsequently. For example, we might choose to draw our original starting triangle with sides of pink, puce and purple (represented by solid, dashed and dotted lines in the diagrams that follow). Then, whenever it subdivided, we would always be able to find a way (but only one way) to draw the new lines so that each newly formed triangle had sides of these three colors.

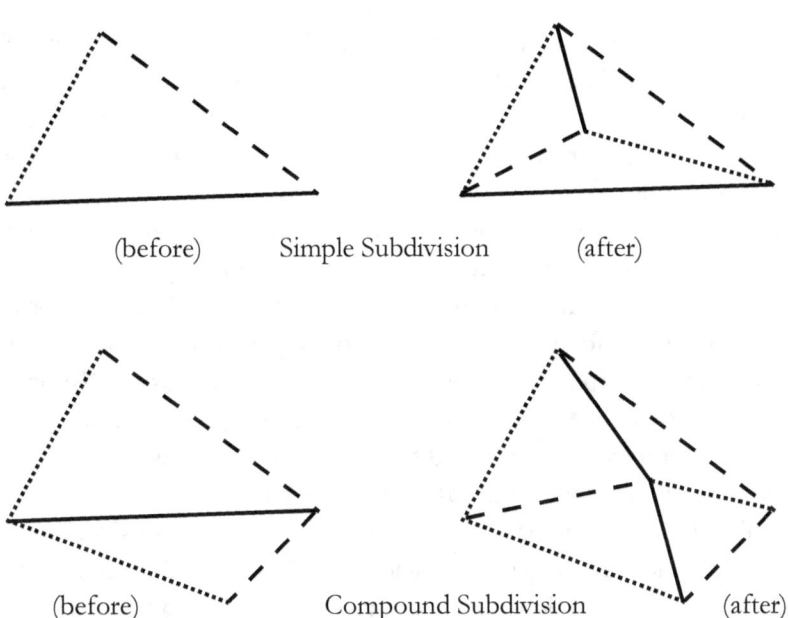

(before)   Simple Subdivision   (after)

(before)   Compound Subdivision   (after)

(In every compound subdivision, the two subdividing triangles' pink, puce and purple edges go in opposite directions: one clockwise, the other counterclockwise.)

The following diagram shows how the initial triangle's coloring relates to the coloring at time 13:

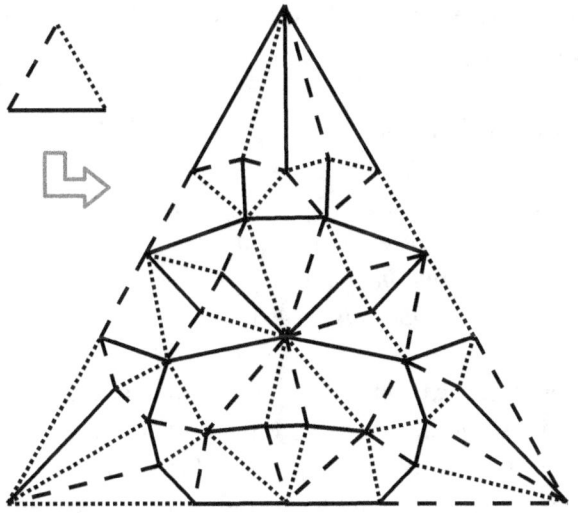

Turning now to the dual graph, we can also color its links to match the corresponding lines in the L2 graph. Then every point in the dual graph would have a pink link, a puce link and a purple link (and no others) meeting at it. For about half of these points, the pink-puce-purple would go in a clockwise direction, and for remainder, the other way. It would be tempting to label these as representing matter and antimatter, but so far I have found no evidence to support this analogy.

However, the dual graph does display a "color confinement" property similar to that found in quantum chromodynamics. This is a theory that treats baryons (particles like protons and neutrons) as consisting of three differently colored quarks (although the colors are called red, green and blue rather than pink, puce and purple). But the principle is the same: the three different colors are always found together.

Of course, the analogy is not exact. We are now considering the points of our dual graph as baryons and the *ends* of the links joining them as quarks, whereas previously we considered the points to be quarks and the links to be the gluons binding them. But when we extend the model into three dimensions, hopefully we will find these two properties (confinement and binding) applying to three elements of the model rather than just the two we have now. This could allow us to identify these elements as baryons, quarks and gluons that would display their properties in a manner similar to the properties of quantum chromodynamics. Specifically, the baryons would consist of three differently colored quarks (color confinement) and the quarks would be bound by gluons (asymptotic freedom).

Effects and their causes

Time present and time past
Are both perhaps present in time future,
And time future contained in time past.
—T.S. Eliot (*Four Quartets*, 1943)

If we rearrange the elements of a Soddy quartet $\{v, w, x, y\}$ into non-decreasing order so that $v \leq w \leq x \leq y$, then we can regard it as being "in existence" from time x to time y, just like the triangle it represents and just like its corresponding point in the dual graph. Then, as I mentioned earlier on, whenever two triangles participate in a compound subdivision, their corresponding Soddy quartets will share the same v, w and y values and will differ only in their x values. These two Soddy quartets can therefore be considered as spatially linked, since their corresponding points are linked in the dual graph and since they are both destroyed at time y.

It turns out that there are other ways in which pairs of Soddy quartets can be linked by having three members in common. This is not surprising, since every Soddy quartet is a solution to the Soddy equation

$$(v + w + x + y)^2 = 3(v^2 + w^2 + x^2 + y^2)$$

which is a quadratic involving four variables. Now if we regard three of the four members v, w, x and y as being fixed and then solve for the fourth, we will in general find two solutions (since we are solving a quadratic equation). And this will be true whether we are solving for v, w, x or y.

This means that in general (excluding the special kissing-sister cases that involve repeated numbers), each Soddy quartet is linked to four others that share three members in common with it. For example, the quartet $\{4, 7, 19, 21\}$ is linked to the four quartets $\{7, 19, 21, 43\}$, $\{4, 19, 21, 37\}$, $\{4, 7, 13, 21\}$ and $\{4, 7, 9, 19\}$, depending on which member is solved for.

Since each quartet "exists" between the times given by its third and fourth members, we can regard the quartet $\{4, 7, 19, 21\}$ as linked to a single temporal predecessor, a single spatial partner, and two temporal successors, as shown in the following diagram:

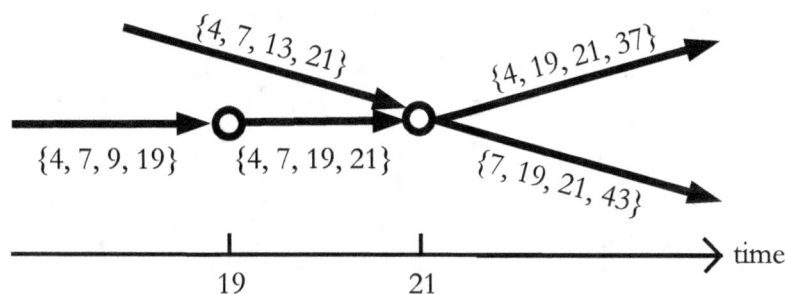

About 77% of Soddy quartets (those involved in compound subdivisions) follow this pattern of having one predecessor, one partner and two successors. (Note that the partner quartet, $\{4, 7, 13, 21\}$ in the diagram, is not linked to either of the two successors; it has its own two successors.) The other 23% of Soddy quartets (which represent simple subdivisions) have one predecessor, no partners, and three successors, as shown in this diagram:

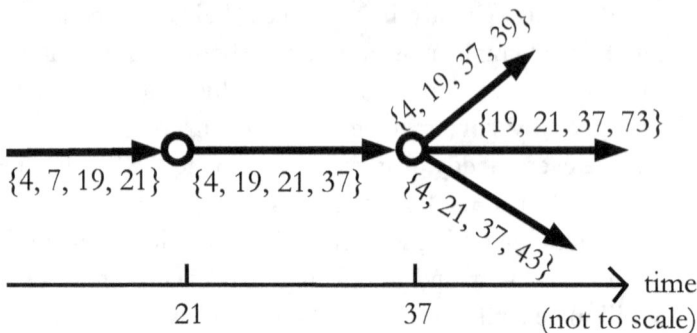

But whichever pattern it follows, *every* Soddy quartet has a unique predecessor, which can be interpreted as a cause-and-effect relationship. Furthermore, in each sector of the L2 model, the two triangles corresponding to any quartet and its predecessor always have some area in common, so there is no "spooky action at a distance" involved in their relationship.

When we trace back the "causal chain" of predecessors for a particular quartet, some interesting patterns emerge. For example, we can trace the pedigree of the quartet $\{7, 817, 876, 973\}$ back to the initial quartet $\{1, 1, 1, 3\}$ through 27 generations (listed here in forward order):

Gen.	Quartet	Duration	Gen.	Quartet	Duration
1	$\{1, 1, 1, 3\}$	2	2	$\{1, 1, 3, 4\}$	1
3	$\{1, 3, 4, 7\}$	3	4	$\{1, 3, 7, 7\}$	0
5	$\{3, 7, 7, 16\}$	9	6	$\{3, 7, 16, 19\}$	3
7	$\{7, 16, 19, 39\}$	20	8	$\{7, 16, 39, 43\}$	4
9	$\{7, 39, 43, 73\}$	30	10	$\{7, 39, 73, 76\}$	3
11	$\{7, 73, 76, 117\}$	41	12	$\{7, 76, 117, 127\}$	10
13	$\{7, 117, 127, 175\}$	48	14	$\{7, 127, 175, 192\}$	17
15	$\{7, 175, 192, 247\}$	55	16	$\{7, 192, 247, 271\}$	24
17	$\{7, 247, 271, 333\}$	62	18	$\{7, 271, 333, 364\}$	31
19	$\{7, 333, 364, 433\}$	69	20	$\{7, 364, 433, 471\}$	38
21	$\{7, 433, 471, 547\}$	76	22	$\{7, 471, 547, 592\}$	45
23	$\{7, 547, 592, 675\}$	83	24	$\{7, 592, 675, 727\}$	52
25	$\{7, 675, 727, 817\}$	90	26	$\{7, 727, 817, 876\}$	59
27	$\{7, 817, 876, 973\}$	97			

Each quartet's "duration" represents the amount of time it exists, calculated as its y value minus its x value. Notice how the durations alternate between long and short, and how, after generation 11, they increase by 7 on each row (which, not coincidentally, is the first number in these quartets).

How to interpret this physically? Well, the particular quartet we selected contains a 7 and therefore is located on the edge of a region of the dual graph that formed at time 7. As the model matures, its oldest regions develop more and more sides and become "white holes" in the space of the dual graph. Here is a portion of the dual graph at time 876 (the final generation in our chart): each of the small circles represents a Soddy quartet, with the larger circle at the bottom edge of the white area representing our selected quartet.

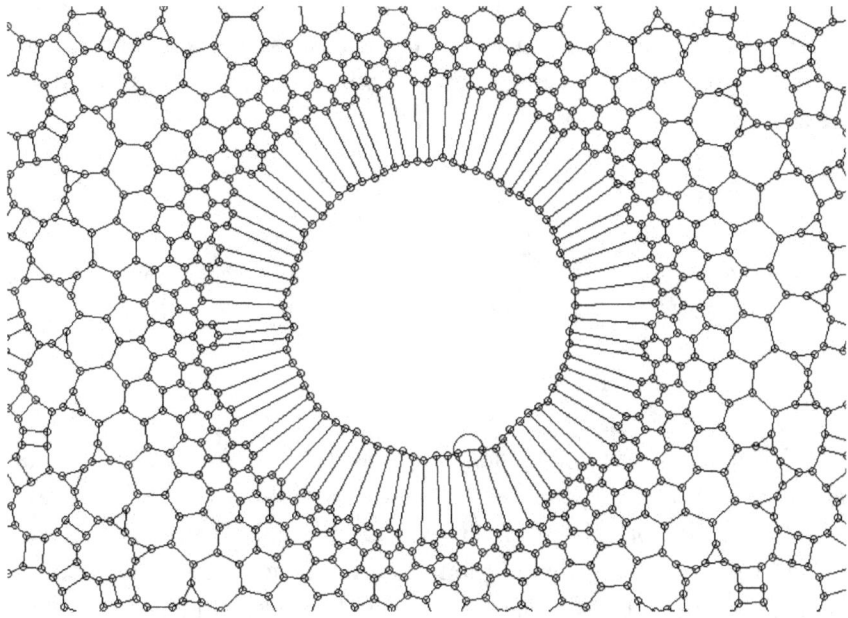

Now, in our dual space, all the links are defined as having length 1, so this picture is not to scale; instead, it is scaled to our geometric model, with each region's area being proportional to the energy it contains. The picture therefore represents a portion of our dual space, a portion which is distorted because it contains a large concentration of energy. In physical terms, this would correspond to a black hole in space (but it's shown as white here for clarity of printing). Like the black holes of physics, this white hole has no internal structure and occupies no space (since it contains no points of the dual graph, its area is zero). Yet its perimeter is 75 units long, so it's a lot bigger on the outside than the inside (a SIDRAT?).

The causal chain of predecessor quartets listed above (from generation 11 onward) can thus perhaps be regarded physically as representing a particle of matter/energy that "falls into" a black hole. At each generation it spins off a successor quartet that is not part of the causal chain, and it is tempting to regard this as representing something like the Hawking radiation which black holes are believed to emit. More significantly, if you consider the alternation of long- and short-duration quartets as a type of vibration, then you will see that (from our perspective) the frequency of this vibration diminishes over time. So if a counter were somehow attached to the particle to measure the number of interactions it experienced, that "clock" would appear to be slowing down from our perspective (although not from the particle's). Meanwhile, the scaled energy in the white hole is growing all the time, so this could be considered as a form of "gravitational time dilation"—which brings me to my next topic.

What is gravity?

If you fall from a high tower, you fall quicker and quicker and quicker; a judicious selection of a tower will ensure any rate of speed.
—Stephen Leacock (1910)

In ancient times, the tendency of objects to fall was believed to be an inherent property of earthbound matter. By Galileo's time, it was realized that heavy objects don't fall any faster than light ones (although they do fall harder). Then Newton formulated his famous law of gravitation, stating that every particle of matter in the Universe attracts every other particle, with a force that varies with the inverse-square of the distance separating them. This force accounted for the motion of the apple, the moon and the planets, but Newton was unable to find a mechanism to explain it satisfactorily. (Modern scientists are also puzzled as to why gravity is so many times weaker than the three other "fundamental forces of nature" they have identified: the strong and weak nuclear forces and the electromagnetic force. I suspect this disparity is due to the great age of our Universe.)

In 1915 Einstein, having realized that gravitational force and acceleration are one and the same thing, produced his general theory of relativity. This explains gravity as a curvature in space-time caused by the presence of matter. One feature of this theory is that clocks run slower in stronger gravitational fields. You can verify this at home by a simple experiment using a bookcase and two atomic clocks. All you have to do is synchronize the two clocks, set them on different shelves and wait a year.

Now, Newton's inverse-square law states that the earth's gravity weakens as you get farther from its center. Accordingly, Einstein's theories predict that the higher clock should run slightly faster than the lower one. And indeed you will find this to be the case. (By the way, general relativity's effects are several times larger than those of special relativity attributable to the Earth's rotation.) When you compare the clocks after a year, the upper clock should be ahead of the lower one by about one nanosecond for each foot of difference in their height. (But remember to stand exactly half-way between the clocks when you compare them, since the speed of light is—coincidentally—also about one foot per nanosecond.)

All this follows from Einstein's general relativity, but I find the idea of a four-dimensional curved space-time somewhat hard to grasp intuitively. (Those often-seen diagrams which show a ball rolling on a rubber sheet distorted by a heavy weight just seem to explain gravity *in terms of itself*. Like the ball's path, the explanation is circular.) So instead I like to consider gravity as a *slowing of time* that occurs near massive objects. And even though this effect is *tiny* at the Earth's surface, and the way it tapers off with increasing height is even tinier, this tapering is still enough to give the acceleration of 32 feet per second per second with which we are all familiar. To show the equivalence, we just have to apply a conversion factor of c^2, the square of the speed of light. (This quantity is central to relativity theory: it is also the conversion factor between matter and energy in the equation $E = Mc^2$.) The approximate calculation goes like this:

Difference in clock speeds (1 nanosecond per year per foot)
× speed of light, squared (1 foot per nanosecond)^2
= 1 foot per year per nanosecond
= 32 feet per second per second.

This slowing of time also corresponds to the "gravitational lensing" effect, which is the bending of light as it passes close by a celestial body like a star. You can consider light as a wavefront, every part of which advances at the speed of light. Since (from our perspective) time runs slower the nearer you are to the star, those parts of the wavefront which are nearest the star will travel less far than those that are more distant, in equal amounts of time. This has the effect of deflecting the wavefront:

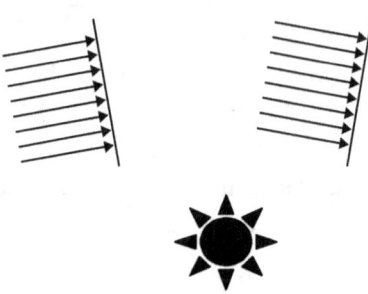

In PQR Theory, the passage of time is the same thing as the subdivision of space, so gravity should occur because *the presence of energy inhibits the subdivision of space.* In other words, space subdivides more slowly in the presence of matter. Or, when we apply our scaling factor, we would say that space expands more rapidly in the absence of matter. Thus, where matter is present, there is a *deficit in the expansion of space* —the space around a large mass expands more slowly than empty space.

In everyday terms, this means that we can consider matter as continually "sucking in" the space that surrounds it, much as a vacuum cleaner nozzle sucks in the surrounding air. Thus, each celestial body seems to create an inward "space wind" that exerts a pressure on neighboring objects, making them accelerate toward it. (The space itself does not move, but it expands more rapidly above the apple than below it, which is what causes the downward force on the apple.) The effects of this space wind would be equally distributed across the surface of any imaginary sphere centered around our vacuum cleaner. (A Dyson sphere?) As a result, this pressure would diminish in proportion to the inverse-square of distance—just like gravity does. And Newton's law of gravitation also implies that equal masses should suck in equal amounts of space in equal times.

The L2 model displays signs of a gravitational process at work, as we previously saw happening near the "white hole." This process is hard to observe directly because of the way space and time are interrelated, but we can see its effects by studying how the distance between pairs of points increases over time. (We have to use the points of the L2 model rather than its links or its triangles, because the links and triangles are continually disappearing.) We define the distance in a way that is consistent with the definition we used for the dual graph: the distance between any two points of L2 at a given model-time is simply the minimum number of intermediate steps needed in order to get from one to the other using adjacent triangles as stepping stones. (Adjacent means sharing an edge, so shortcuts across corners are not permitted.)

If we consider a pair of points A, B in the model that are a given distance d_1 apart at a certain model-time t_1, we can then recalculate their separation d_2 at some later model-time t_2, to see if this is influenced by the energies associated with the points A and B. (We define the energy at a given point and time as one-third of the total area of the then-existing triangles that meet there. Like our previous definitions, this gives constant total energy of 1 in the model, which—as before—we then scale up by t^2, the square of the time.)

Now, Newton's laws imply that for given values of t_1, t_2 and d_1, the final separation d_2 will depend linearly on the sum of the two energies at A and B. But when we calculate this in practice, we find that the result for any individual pair appears to fluctuate unpredictably. (In a physical sense, we can attribute this to the "random" quantum nature of the processes that make space subdivide.) This shows that our space wind does not blow uniformly. But then, neither does an atmospheric wind, which consists of many air molecules moving mostly at random but with a small overall trend. (It only requires a tiny imbalance in the random motion—less than one part in a thousand—to create a stiff breeze.)

So instead of considering individual pairs of points, we consider *all* pairs of points that exist in the model with separation d_1 at time t_1 and then plot their *average* separation at a later time t_2. And we do indeed find this average to be linearly correlated with the sums of the point energies, as shown by the bubble-scatter diagram on the next page. (For this particular diagram, t_1 was chosen as 100, t_2 as 400 and d_1 as 20, but a similar pattern would appear with any other choice of values that gives a fair sampling.)

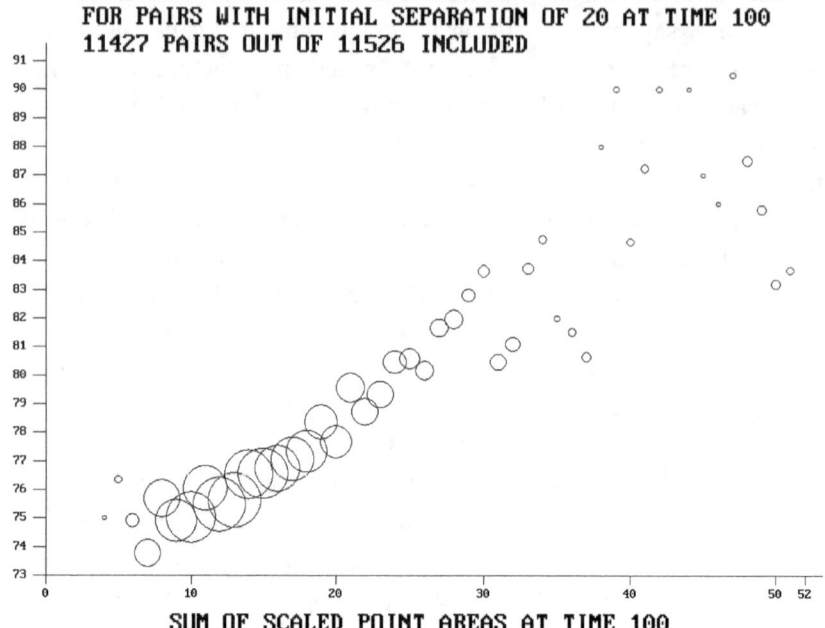

**L2 MODEL: AVERAGE POINT-PAIR SEPARATIONS AT TIME 400
FOR PAIRS WITH INITIAL SEPARATION OF 20 AT TIME 100
11427 PAIRS OUT OF 11526 INCLUDED**

SUM OF SCALED POINT AREAS AT TIME 100

This diagram was compiled by finding all pairs of points in the L2 model which had a separation of 20 at time 100; there were 11,526 such pairs. For each pair, the scaled areas of its two points at time 100 were added together and the sum rounded to the nearest integer. This number represents the total energy initially contained in the pair. The separation of each pair was then recalculated at time 400 and averaged over all pairs with the same energy value. The results are plotted above; the area of each bubble represents the number of pairs with each initial energy value. To enhance its clarity, the chart only includes pairs with energies up to 52 (this limit was chosen so as to include at least 99% of the pairs). So some small outlying bubbles are cut off on the right-hand side of the plot, but they fall into the same pattern.

The chart's trend is clear: the linear correlation between separation and energy is evident. The only problem is *it goes the wrong way*. The pairs that (on average) separate farther are the ones with higher energies, rather than the ones with lower energies as Newton predicted. So gravity appears to be a *repulsive* force in this model (it should accordingly be renamed *levity)*. Newton's gravity sucks, but this one blows.

Joking aside, if this correlation is not reversed when we add an extra dimension and investigate L3, then it will be necessary to revise our model, or at least our physical interpretation of it. But hopefully, when L3 is investigated, it will turn out that the regions surrounding higher-energy points do indeed tend to subdivide more slowly as the model develops than those surrounding lower-energy points. In terms of our physical interpretation, this would indeed mean that the presence of matter was slowing the subdivision of space. Then, when we apply our scaling factor to get an expanding universe, it would seem as if *matter eats space and slows time.* Which—thanks to Newton and Einstein—we know is why apples fall and satellites stay in orbit.

Fun with photons

For the following thought experiment we're going to need some polaroid. No, not those retro instant pictures, but the semitransparent plastic material used in sunglass lenses. (Don't use those 3-D glasses from the cinema: most work on a different system nowadays.) Otherwise known as polarizing filters, these lenses have some interesting properties.

Without going into too much detail, the plastic in these lenses contains long-chain molecules arranged in parallel lines; these are designed to absorb the horizontally vibrating component of light waves while allowing the vertical portion to pass. When sunlight shines through one of these filters, about half of it is absorbed and the other half gets through. The light is said to be *vertically polarized* after it passes through the lens, which is taken to mean that its waves vibrate up and down but not from side to side. (If you were to rotate the lens through 90°, the light would instead be polarized horizontally. And if you rotated it through an additional 90°, its polarization would be back to vertical again.) In general, polarizing filters have a *direction of polarization*: they are said to transmit waves vibrating in this direction but block vibrations at right angles to it.

You can check this out by shining a flashlight beam through a sunglass lens and then placing a second lens after the first one: if its polarizing direction is parallel to the first, it will hardly block any of the light, but if you turn it through 90°, then it will block all the light, as the following diagram illustrates:

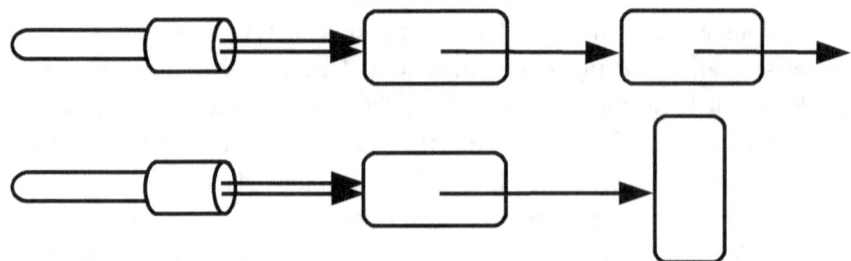

(Incidentally, if you were to put a *third* filter in between the two lenses and turn it to 45°, you would find some light passing through the system again. This filter "twists" the polarization of the light passing through it, which allows some to get through the last filter also.)

Returning to the two-lens setup illustrated above, if you turn the second lens to some intermediate angle between 0 and 90°, you will find that a proportion of the light passes through it. And since light consists of individual photons, we can in principle conduct this experiment using a very dim light source that only sends one photon at a time through the first filter. When we do this, we find that the process which determines whether a photon passes through the second filter or gets absorbed by it appears to be totally random. However, by repeating the experiment many times, we can statistically estimate the probability of transmission through the second filter, and we find that this depends on the angle α between the polarizing directions of the two filters. With the filters aligned ($\alpha = 0$), the probability of transmission is 1; when the filters are at right angles ($\alpha = 90°$), the probability of transmission is 0; with the filters at 45°, the probability of transmission is ½. In general, the chance that a single photon will pass through the second filter is $\cos^2 \alpha$ (the square of the cosine of the angle between the filters), and this is also the proportion of light that is transmitted when we use a more powerful light source.

Now, we said before that the photons leaving the first filter were all polarized vertically, but do we know this for sure? All the experiment tells us is whether a particular photon was absorbed or transmitted. No matter how we set things up, we can only ever extract a single bit of information out of any one photon. There is no way to measure the polarization of a single photon more accurately than to say "well, it passed through a filter oriented at x degrees, so it certainly wasn't polarized exactly at right angles to that; it was *most likely* polarized somewhere around x degrees, but it could have been almost 90 degrees out on either side."

We normally say that a photon which has passed through a vertical filter is *definitely* polarized vertically, and that it is *randomly* transmitted by the second filter. But it is just as meaningful to say that it is *randomly* polarized at an angle of up to 45° on either side of vertical, and that when it hits the second filter it will be *definitely* transmitted if the angle between its polarization and the second filter is less than 45° and *definitely* absorbed if it is more than that. If the probability distribution for the polarization angle θ is given by the formula

$$p(\theta) = \cos(2\theta) \qquad\qquad (-45° < \theta < 45°)$$

then this alternative model gives the same probability of transmission at the second filter as we had before ($\cos^2 \alpha$).

Which raises the question, do the photons leaving the first filter really *have* a definite angle of polarization? Or is this merely an illusion, a convenient model to explain what we see? It seems to me that the photon's polarization angle is *not* well defined, and that the photon only carries a single bit of information about its orientation. (In quantum physics, this would correspond to its helicity, a quantity which takes one of only two possible values: +1 or −1.)

How then does the photon "know" how to behave when it hits the second filter? My answer is that it doesn't need to carry any information, since its travel from the first filter to the second is instantaneous. From the photon's perspective, it is simultaneously at *both* filters at once, at which instant it can "measure" the angle between them and "decide" whether or not to pass through the second filter.

A similar reasoning also accounts for the famous double-slit experiment, in which photons (or more massive particles) are fired at a screen through a barrier containing two parallel slits. (A variant of this experiment uses half-silvered mirrors to send photons down alternate paths to detectors.) In either case, there is no way of telling which route any particular particle has taken. The particle doesn't know or care which route it takes, because it passes from its source to its destination in a single step of its existence (or, equivalently, in a chain of linked steps which do not cause any external effects or receive any external influences). Although you know it passed through the barrier, it would be meaningless to ask it "which slit did you go through?" You might equally well ask it "which part of the barrier did you fail to hit?"

In terms of the L2 model, we can now relate this to our discussion of Soddy quartets earlier in this chapter. For example, we saw how the quartet {4, 19, 21, 37} was "in existence" from time 21 to time 37. It "spanned the gap" between an interaction at time 21 and another at time 37 but had no interactions in the intervening period. And this, I believe, is how photons (and more massive particles) operate in our Universe. They jump from one position in our granular space-time to the next, without needing to pass through any intermediate granules. And in between these interactions, they are of course undetectable: they can only be detected when they interact with our measuring apparatus. This, I think, may also explain why photons don't go through solid walls: each Soddy quartet can only span a small expanse of space in our model, which suggests that a photon would not be able to pass through a barrier of any substantial thickness without encountering some interference from solid matter that would break the influence-free chain of transmission. (But I haven't yet worked out exactly how this principle might operate in a three-dimensional model.)

8. Level 3: The Four-D Froth

Life is mostly froth and bubble.
—Adam Lindsay Gordon (1866)

The principles used to construct the one- and two-dimensional L1 and L2 models can be extended to make a model with three dimensions of space, plus one time dimension, known as the L3 model or Four-D Froth. Each of our previous construction methods offers a natural way of adding this extra dimension. And as before, all the methods result in the same structure in three dimensions: a tetrahedral space divided into polyhedral regions. Unfortunately, the thinness of these pages precludes proper pictures, so the description of this model has to be mostly verbal.

However, you can visualize the general form of this structure at home: half-fill the kitchen sink with warm water and add some dish soap. (Farey liquid?) Blow into the water through a straw and let the bubbles pile up on the surface. Notice the shape of the bubbles that fill the sink: apart from the top layer, they are all polyhedral (spherical bubbles wouldn't fill the space). This froth is similar to the interior of the L3 model, whose space is divided into an ever-increasing number of polyhedral regions, much as the L2 model was divided into triangles. The analogy is not exact, however, since the new bubbles all appear at the same place in the sink, while the spaces of the Four-D Froth subdivide throughout the model.

Geometrical construction: Ford hyperspheres

Like the previous models, the Level Three model can be built geometrically, this time as a system of Ford hyperspheres. Although it is harder to visualize, the construction follows the exact same principles as before.

In Level One, we used a two-dimensional plane and started with two circles resting on a straight one-dimensional line embedded in the plane. On Level Two, we used a three-dimensional space and started with three spheres touching a flat two-dimensional plane embedded in the space. For Level Three, we use a four-dimensional hyperspace and start with four hyperspheres that touch a three-dimensional Euclidean space embedded in the hyperspace. (All spaces used in the geometrical constructions are Euclidean or "flat," meaning that the usual rules of geometry apply.)

As in the previous levels, these four hyperspheres are all of equal size, and they all touch each other as well as touching the embedded three-dimensional "subspace." The four points at which they touch the subspace are all equidistant from each other, so they form the vertices of a regular tetrahedron in that three-dimensional space, and this tetrahedron will become the three-dimensional space of our model. This picture shows the boundaries of that space and the four points (marked by circles) where our hyperspheres touch it:

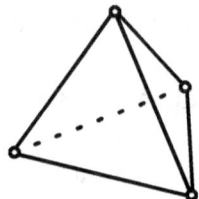

The picture doesn't show the remainder of the hyperspheres because they are not in the three-dimensional space it represents (and also because I don't know how to draw them).

Now, as before, we keep adding hyperspheres in our four-dimensional hyperspace, with each new hypersphere touching a single point of our tetrahedral subspace. In this way, we fill the gaps between that subspace and the existing hyperspheres, and sooner or later every rational point in our tetrahedron will have a hypersphere touching it. As before, these hyperspheres don't overlap, and their diameters are all exact fractions of the original hyperspheres' diameters. And as before, the order of each hypersphere is the reciprocal of its diameter and denotes the "time" at which it is created. In this model, there are hyperspheres of every order except 2, 5, 10 and 14. So, unlike the previous models, after the first few stages there are no "gaps in time" where nothing happens.

While the L1 model had two-way symmetry and the L2 model had six-way symmetry, the L3 model has twenty-four-way symmetry. (In general, the Level n model has $(n+1)!$-way symmetry, corresponding to the number of ways of arranging the elements of a duad, triad or tetrad.) In other words, the L3 model can be sliced into twenty-four equal segments that meet at its center. Each segment is an irregular tetrahedron, and although they are all of the same size, they are of two different types: there are twelve identical "left-handed" segments and twelve identical "right-handed" segments that are mirror images of the first twelve.

In the assembled model, the two types of segments alternate: each left-handed segment has three of its faces touching a right-handed segment, and vice versa. The fourth face of each segment is on the outside of the tetrahedron, as shown here.

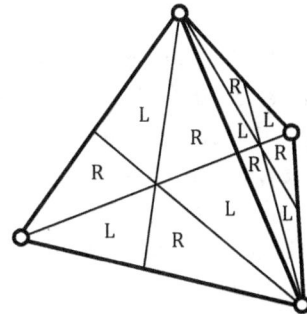

As you can see, each face of the tetrahedron is a copy of the L2 model, and each edge of the tetrahedron, along with being an edge of two such copies, is also a copy of the L1 model.

The model's symmetry persists through time, so that everything that happens in one segment happens in all the others. (If you were living in this model, there would also be eleven identical copies of you and twelve mirror images of you, symmetrically distributed around this universe, one per segment. The boundaries between the segments would appear to be mirrors, so if you had Marxist leanings you and your counterparts could reenact that famous scene from *Duck Soup*—with twelve Grouchos and twelve Harpos in twelve places at once.)

As the model develops, new points are continually being created. Each corresponds to a new hypersphere that "fills a gap" in our four-dimensional space by touching a number of previously created hyperspheres. In the L2 model, each newly created sphere only touched three or four of its predecessors, but in this model the hyperspheres have four to eight parents. And whereas in the L2 model there were only two types of subdivision, in this model the structure is far more varied and complex. If we sort each hypersphere's parents by order and examine how they touch each other, we find twenty-five different possible permutations occurring in the interior of the model. (There are also five cut-down versions of these permutations that only occur on the faces of the model.) Some of these permutations are rearrangements of each other, but even when we disregard the order among parents, there are still nine different types of subdivision that occur in the model.

Of course when we look at our three-dimensional subspace, these hyperspheres only appear as points. But, like in the previous levels, we can draw lines in the 3-D subspace joining pairs of points whose hyperspheres touch each other in the four-dimensional space. Then, when a new point forms in our subspace, it forms within the polyhedral volume bounded by the lines that connect its parent points. The faces of these polyhedra are triangles and plane quadrilaterals. (Correspondingly, in the L2 model, new points were formed within triangles (simple subdivision) and quadrilaterals (compound subdivision).)

In the L3 model, the nine different types of subdivisions appear to be associated with nine different polyhedral shapes, as follows:

(A) a tetrahedron; (B) a tetrahedron with a triangular flap attached to one edge; (C) a pyramid; (D) two tetrahedra hinged together (sharing an edge in common); (E) a pyramid joined to a tetrahedron (sharing a face in common); (F) an octahedron (two pyramids joined base to base); (G) a pyramid joined to two tetrahedra; (H) an octahedron joined to a tetrahedron; and (I) an octahedron joined to two tetrahedra.

These pyramids all have flat (but not necessarily square) four-sided bases, and although the octahedra aren't symmetrical, each one's three diagonals (the lines joining opposite pairs of points) meet in a single point. As a result, each octahedron can be sliced into two pyramids in three different ways. This is illustrated in the picture of a (type F) octahedron below: the lines AF, BE and CD all meet in a common point X in the interior, and each of the three circuits ABFE, ACFD and BCED lies in a single plane that divides the octahedron into two pyramids.

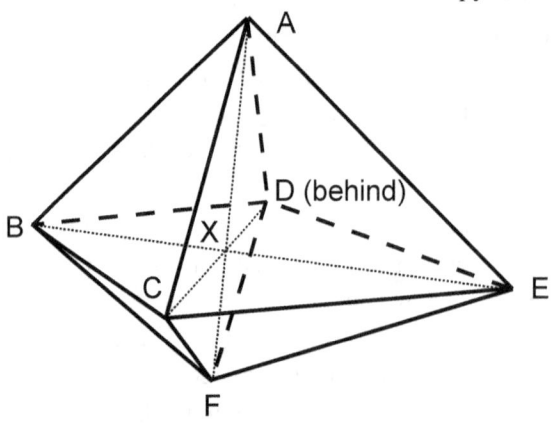

Also, where two tetrahedra share a common edge (so that they have two vertices in common), their remaining four vertices also lie in a single plane, as illustrated in the picture (of a type D shape) below: the two tetrahedra ABCD and ABEF share the edge AB, and the vertices C, D, E and F all lie in a single plane.

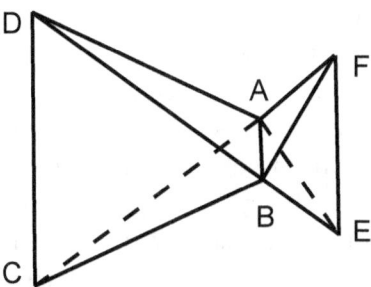

It may well be possible to simplify these shapes, since some of their lines presumably get destroyed by subdivisions as happened in the L2 model. This is an area which I have not yet fully explored, so it remains to be seen whether these shapes have any connection with the "ampli-tuhedra" that have recently been postulated by quantum field theorists.

However, it does appear that the space of the L3 model is subdi-vided into shapes made up of pyramids and tetrahedra joined together in various ways. And since a pyramid can be sliced diagonally into two tetrahedra, we can calculate the volume of these shapes by adding the volume of the constituent tetrahedra.

The following table shows how the number of points in this model increases over time:

Order or time (t)	No. of points (n)	n/t^3	Order or time (t)	No. of points (n)	n/t^3
1	4	4	15	175	0.051852
2	4	0.5	20	439	0.054875
3	8	0.29630	30	1,187	0.043963
4	14	0.21875	40	2,547	0.039797
5	14	0.112	50	5,143	0.041144
6	18	0.08333	100	36,917	0.036917
7	42	0.12245	250	551,491	0.035295
8	43	0.08398	500	4,347,057	0.034776
9	67	0.09191	1,000	34,472,293	0.034472
10	67	0.067	2,000	274,747,179	0.034343

Over the long term, the number of points settles down to approximately 0.0343 times the cube of the order. (Curiously enough, this number is close to $3\pi/2$ times α, the mysterious fine structure constant of physics, which comes out at approximately 0.03438796, but that may well be just a coincidence.)

The graph-theoretical aspects of the L2 model can also readily be extended to L3 in various ways (although the lines would cross if you tried to plot them on paper). One such graph would plot all points that existed at a certain time, with links to connect those whose hyperspheres touched in the fourth dimension. By analogy with the previous model, these links would presumably get destroyed eventually.

A second way to plot a graph would be to make its points correspond with the polyhedral spaces of L3, with links connecting polyhedra that are adjacent (which now means sharing a face in common). A third way would be to plot a graph whose points correspond to the faces of the polyhedral spaces, with links to connect pairs of faces that meet at an edge. Yet another way (but probably not a very useful one) would be to plot a graph whose points correspond to the links of the first graph mentioned above, and whose links correspond to its vertices. These graphs could be used as a way of defining distances between various parts of the model, as we saw done for L2, but this is an area I have not researched in any detail.

Numerical construction: Tetrads

Similarly to the triads that constructed our L2 model, we can also build the L3 model using *tetrads*. A tetrad is defined by four coprime non-negative integers a, b, c and d, and written **a:b:c:d>y**. If we let $m = a + b + c + d$ then y, the *order* of the tetrad, is given by the formula

$$y = \frac{m^2}{\gcd\left(\begin{array}{l} m^2,\ a^2+b^2+c^2+ab+ac+bc,\ a^2+b^2+d^2+ab+ad+bd, \\ a^2+c^2+d^2+ac+ad+cd,\ b^2+c^2+d^2+bc+bd+cd \end{array}\right)}$$

where gcd denotes the greatest common divisor, as before.

And as before, each tetrad corresponds with exactly one point in our geometric construction, and vice versa, since y is the order of the point with barycentric coordinates $a:b:c:d$.

Note that the order of the four numbers that define a tetrad is significant. If they are all different, then they have twenty-four different possible arrangements, making twenty-four different tetrads that correspond to twenty-four symmetrically distributed points, one in each segment of the model. If two of the four numbers are the same, then they have only twelve possible arrangements and generate two symmetrically matching points on each of the six planes of symmetry of our regular tetrahedron. If three of the four numbers are the same, then they only have four possible arrangements, and they generate four symmetrically matching points, one on each axis of our tetrahedron. If the tetrad contains two different numbers, each of which occurs twice, then there are six possible arrangements, and they generate two symmetrically matching points on each of the three lines joining the midpoints of opposite edges of the tetrahedron. Finally, if all four numbers are the same (in which case they all have to be 1, or they would not be coprime), they generate the tetrad **1:1:1:1>8**, which corresponds to the center of the tetrahedron.

Note also that one, two or three of the numbers in a tetrad (but not all four) may be zero. Tetrads that contain a single zero correspond to points on a face of the tetrahedron; there are four possible positions for the zero, corresponding to the four faces of the tetrahedron. Tetrads that contain two zeros correspond to points along an edge; there are six possible combinations of positions for the zeros, which correspond to the six edges of the tetrahedron. Tetrads that contain three zeros (in which case the fourth number has to be 1) correspond to the four corners of the tetrahedron. The following diagram shows how some of this works:

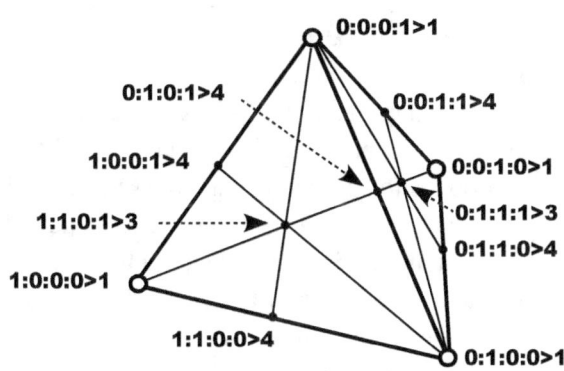

Algebraic construction: Gosset quintets

In n-ic space the kissing pairs
Are hyperspheres, and Truth declares—
As n + 2 such osculate
Each with an n + 1 fold mate
The square of the sum of all the bends
Is n times the sum of their squares.
—Thorold Gosset (1937)

These additional lines by Thorold Gosset generalized the theorem of "The Kiss Precise" to the case of four or more dimensions, where $n + 2$ hyperspheres all touch each other in an n-dimensional space. Since the flat 3-D subspace of our model is equivalent to a hypersphere with zero bend, we can apply Gosset's formula to our Four-D Froth by setting $n = 4$. The poem then tells us that where five Ford hyperspheres all touch each other (and our subspace), their bends, and thus their orders u, v, w, x and y, are related by the equation

$$(u + v + w + x + y)^2 = 4(u^2 + v^2 + w^2 + x^2 + y^2).$$

I call any five coprime Natural numbers that satisfy this equation a *Gosset quintet*. For example, the set $\{3, 4, 7, 9, 13\}$ satisfies this equation, since when you plug in these numbers, both sides of the equation come out at 1,296.

Clearly, whenever a point of order y forms in the 3-D space, inside a tetrahedron whose vertices have orders u, v, w and x, this equation will be satisfied (since the five hyperspheres involved all touch each other). What is less clear is how every Gosset quintet can be made to correspond with a subdividing space in the L3 model, especially since (as outlined above) not all of these spaces are tetrahedra.

However, the planar quadrilaterals that form the faces of the non-tetrahedral spaces all have the property that when you add the orders of any pair of diagonally opposite corners, you get the same sum. For example, you might find that a certain quadrilateral's vertices had orders of 16, 19, 38 and 35 as you went around it, and you would notice that $16 + 38 = 19 + 35$. This relationship also extends to the octahedral spaces in the model: each such space has three pairs of diagonally opposite vertices, and the two orders of each pair all add to the same total. In our example octahedron, (shown below, labeled with its corner tetrads) the orders of vertices A and F add up to 46, and so do the orders of vertices B and E and vertices C and D. And the central point X has barycentric coordinates 35:1:3:7, which also add up to 46. (The coordinates of X are equal to those of B and E added together, or of C and D, or of A and twice F. However, when our octahedron subdivides at time 45, the new hypersphere appears not at X, but close by at 34:1:3:7, corresponding to the tetrad **34:1:3:7>45**. Of course, since X is a rational point, a hypersphere will eventually form there, but this does not occur until time 1058, corresponding to the tetrad **35:1:3:7>1058**.)

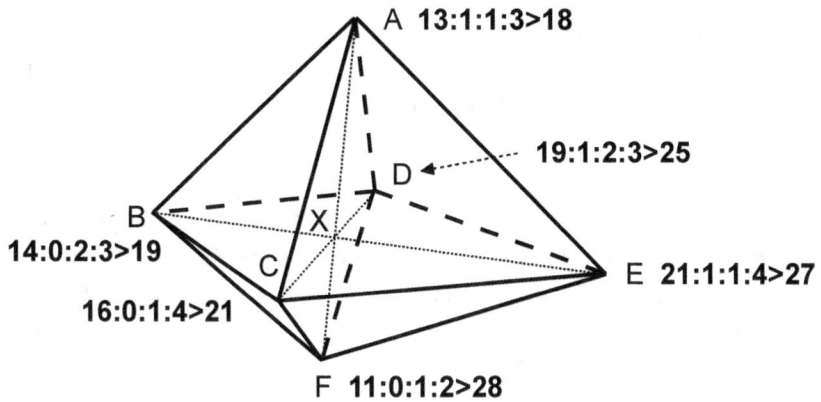

Clearly there is a numerical relationship at play here, but how it ties into the Gosset formula will require some further investigation before I can supply an analysis analogous to the one for Soddy quartets in the L2 model.

Although the total area represented by Soddy quartets came out at exactly the area of the unit triangle in the L2 model, the volume attributable to Gosset quintets in the L3 model appears to be only about 98% of the volume of the unit tetrahedron. The missing 2% appears to be attributable to gaps between the tetrahedral spaces of the model, and these gaps may well correspond to links or faces that are destroyed during the process of subdivision. Hopefully the situation will become clearer once I obtain a better understanding of which links and faces are destroyed and when, and how the pyramidal and octahedral spaces can be related to the Gosset formula. It may prove necessary to adjust the formula or even to include a supplementary equation—or it may simply be a matter of finding the right physical interpretation to reconcile the discrepancy.

For example, it may turn out that the space formed by Gosset quintets (corresponding to one segment of the geometric model) can be detached from that model and analyzed independently. This would eliminate the mirror-image phenomenon that I mentioned earlier and could also allow the model to avoid the "luminiferous aether trap"—the problem that arises if you assume that light propagates through a geometrical framework. Under that assumption its speed, measured experimentally, would appear to vary depending on the experimenter's motion through the framework. The Michelson–Morley experiment showed this not to be the case. Instead, space-time is held to be relativistically invariant, meaning that experimental results should be the same for any observer who is unencumbered by gravity. A model using an algebraic space might be devised to include a Lorentzian metric like those found in relativity theory, in which case it should display the desired invariance.

For example, each Gosset quintet can, in general, be linked to five others that each have four values in common with it, in much the same way that Soddy quartets are linked. These relationships could perhaps be used to devise a space, independent of the geometric model, which incorporates a system of distance measurement that produces space-like and time-like intervals, as in relativity theory. But that is a task for the future.

Future research

As I said at the start of this book, PQR Theory is still a work in progress. I do not yet claim that its models provide an accurate description of our Universe, and its predictions are for the most part as yet untested. That's why it's called a theory.

Before it can justifiably be said that any numerical model (whether it be L3, an L3 variant or a completely different formulation) represents our Universe, it will have to convincingly display features such as momentum, electromagnetism (which includes the propagation of light) and the structure of the atom. It will also need to show relativistic invariance (experimental results must be independent of the motion of the experimenter through space) and match the observations of cosmology. I am confident that with time, patience and perseverance, all these things can be accomplished. In the meantime, if this book stimulates discussion and further research by myself or others, then it will have fulfilled its purpose.

Afterword: Do other universes exist?

2B, or not 2B? That is the question.
—Man waiting at bus stop

It depends upon what the meaning of the word "is" is.
—Bill Clinton (1998)

In the preceding pages, I have discussed several model universes, each defined by a single simple formula using the Natural numbers. And I believe our physical Universe also to be defined by such a formula, although not necessarily by one that I have discussed. So given that various different universes can be constructed mathematically from the Natural numbers, this naturally raises the question *do these other universes exist?*

This turns out to be more a question of philosophy and semantics than one of physics and mathematics. Because if we define our Universe as "everything that exists," then clearly no other universe can exist, or it would be a part of ours. So in our Universe, these other worlds could only exist as mathematical abstractions, rather like the uncalculated digits of π—at least, until someone went and simulated them on a computer.

And yet, if there is a formula that governs our Universe, there may very well be other formulas that also produce some kind of sentient life. These life forms would exist in their own universes, but they could not exist in ours. They might even include philosophers, who would probably conclude that we don't exist—at least not in their universe. But they would consider themselves to be just as real as we consider ourselves.

So which of these two universes would be the "real" one? The answer, of course, depends on where you are standing. Since there is no way to travel to the other universe or communicate with its inhabitants, for all practical purposes you can consider it not to exist. And this will be the case, whichever universe you happen to find yourself in.

The inevitable conclusion is that, like time and space, *reality itself* is a relative phenomenon: one alien's reality is another's science fiction. So the question *do other universes exist?* turns out to be paradoxical. Yes, they exist in their own dimensions, but not in ours. They are just as real as our own Universe—and just as imaginary, too.

GLOSSARY OF SELECTED TERMS

Binary: A number system (base 2) that only uses the digits 0 and 1, instead of 0 through 9. In this system, 10 represents *two*, 100 represents *four*, 1000 represents *eight*, and so on.

Bit: The basic unit of information, a binary digit that can take the value 0 or 1.

Cartesian: relating to the French philosopher and mathematician René Descartes or his system of rectangular coordinates. The principles of modern philosophy were postulated by Descartes. (*Discarding everything he wasn't certain of, he said, "I think, therefore I am a rhubarb tart."*—John Cleese (1966))

Complex numbers: Numbers that consist of a Real number plus an Imaginary number.

Coordinates: A set of numbers that specify a point's position in a space.

Coprime: Describes a group of two or more numbers that share no common divisor greater than 1. For example, 6, 10 and 15 are coprime, as no number (apart from 1) divides evenly into all three. (But 10 and 15 are not coprime, as they are both divisible by 5.)

Curvature: The reciprocal of the radius (1/radius) of a circle, sphere or hypersphere. Straight lines, flat planes and Euclidean (normal) space all have curvatures of zero.

Diameter: The distance across a circle, sphere or hypersphere. Twice the radius.

Dimension: One of the measurements needed in order to specify a position in space or in space-time. Normal space has three dimensions, since a position can be specified using three numbers; time can be considered as a fourth dimension.

Diophantine equation: An equation that has to be solved using whole numbers.

Divisor: A number which divides evenly into another, leaving no remainder. For example, 6 is a divisor of 18, but not of 19. Also known as a *factor.*

Duad: An ordered pair of coprime Natural numbers. The duad defined by the Natural numbers a and b (in that order) is written as **a:b>y**, where **y** denotes the *order* of the duad.

Euclidean: Relating to the geometrical theory of flat spaces postulated by the ancient Greek mathematician Euclid (as opposed to the curved spaces used by Einstein). A Euclidean space may have 1, 2, 3 or more dimensions (respectively a line, plane, normal space or hyperspace).

Finite: Describes something that does not go on forever, but has a definite end.

Hypersphere: The equivalent of a sphere in four dimensions (or more).

Imaginary numbers: A class of artificial numbers, defined as the square roots of negative numbers.

Infinite: Describes something that goes on forever.

Integers: The whole numbers, including negative numbers and zero.

Irrational numbers: Those numbers like π and $\sqrt{2}$ which *cannot* be expressed as the ratio of two integers. When written as a decimal, an Irrational number goes on indefinitely without ever stopping or going into a repeating loop.

Löschian numbers: Numbers of the form $x^2 + y^2 + xy$, where x and y are integers.

Lowest terms: Describes a fraction after any common factors have been divided out from its top and bottom. Fractions are normally written in this manner: for example, we write 2/3 instead of 4/6.

Nanosecond: One billionth (10^{-9}) of a second.

Natural numbers: The "counting numbers," 1, 2, 3 and so on.

Order: The "time" at which an element of the L1, L2 or L3 models is generated.

Photon: A light quantum, in more ways than one: as well as being the fundamental unit of light, it has no mass.

Polygon: A (non-flying) plane shape bounded by straight edges. (Or a non-flying Norwegian Blue.)

Polyhedron (adjective: **Polyhedral**): A solid body (such as a cube or a pyramid) having flat faces, which in turn are bounded by straight edges. (Thus its faces are polygons.)

Primes: Those Natural numbers which cannot be obtained by multiplying two others together: 2, 3, 5, 7, 11, 13, 17, 19 and so on. There are infinitely many primes. (The number 1 is excluded as being a special case.)

Quadratic equation: One that involves squares, but no higher powers.

Quantum (plural: **Quanta**): The smallest possible unit of something. Something that only comes in these particular amounts (like light of a given color, or electric charge) is said to be *quantized*.

Radius: The distance from the center to the edge of a circle, sphere or hypersphere. Half the diameter.

Rational numbers: Those numbers which are the *ratio* of two integers. The set of Rational numbers consists of the integers together with all possible fractions. When written as a decimal, any Rational number will eventually terminate or go into a repeating loop.

Real numbers: The Rational numbers together with the Irrational ones.

Reciprocal: The result of putting 1 over a given number: for example, the reciprocal of 4 is ¼. (As their name implies, reciprocals also work in reverse: the reciprocal of ¼ is 4.)

Space-time: Space and time considered as a unified four-dimensional fabric having three dimensions of space, and time as a fourth dimension.

Tetrad: An ordered set of four coprime Natural numbers. The tetrad defined by the Natural numbers a, b, c and d (in that order) is written as **a:b:c:d>y,** where **y** denotes the *order* of the tetrad.

Tetrahedron (adjective: **Tetrahedral**): A solid body with four triangular faces and four corners. Similar to a pyramid, but with a triangular base instead of a square one.

Triad: An ordered trio of coprime Natural numbers. The triad defined by the Natural numbers a, b and c (in that order) is written as **a:b:c>y,** where **y** denotes the *order* of the triad.

Vertex: A corner of a polyhedron where three or more edges meet.

ABOUT THE AUTHOR

Nicholas Mitchell was born in London, England, and attended Westminster School and Trinity College, Cambridge, where he was a Wrangler and received an MA in mathematics. He worked as a life insurance actuary for a number of years before moving to California, where he became a law office manager. He enjoys creating and solving puzzles, and has written puzzles for British Mensa and the MIT Mystery Hunt.

He lives in Los Angeles with his wife, Dr. Dorine Kramer; they have two grown children and a dog (but no cats, dead or alive).

His Erdös number is 3.

www.ingramcontent.com/pod-product-compliance
Lightning Source LLC
Chambersburg PA
CBHW071315220526
45468CB00001B/376